Numanities - Arts and Humanities in Progress

Volume 6

Series editor

Dario Martinelli, Faculty of Creative Industries, Vilnius Gediminas Technical University, Vilnius, Lithuania

The series originates from the need to create a more proactive platform in the form of monographs and edited volumes in thematic collections, to discuss the current crisis of the humanities and its possible solutions, in a spirit that should be both critical and self-critical.

"Numanities" (New Humanities) aim to unify the various approaches and potentials of the humanities in the context, dynamics and problems of current societies, and in the attempt to overcome the crisis.

The series is intended to target an academic audience interested in the following areas:

- Traditional fields of humanities whose research paths are focused on issues of current concern;
- New fields of humanities emerged to meet the demands of societal changes;
- Multi/Inter/Cross/Transdisciplinary dialogues between humanities and social and/or natural sciences;
- Humanities "in disguise", that is, those fields (currently belonging to other spheres), that remain rooted in a humanistic vision of the world;
- Forms of investigations and reflections, in which the humanities monitor and critically assess their scientific status and social condition;
- Forms of research animated by creative and innovative humanities-based approaches;
- Applied humanities.

More information about this series at http://www.springer.com/series/14105

Alin Olteanu · Andrew Stables
Dumitru Borţun
Editors

Meanings & Co.

The Interdisciplinarity of Communication, Semiotics and Multimodality

Springer

Editors
Alin Olteanu
International Semiotics Institute
Kaunas University of Technology
Kaunas
Lithuania

Andrew Stables
International Semiotics Institute
Kaunas University of Technology
Kaunas
Lithuania

Dumitru Borţun
Faculty of Communication and Public Relations
National University of Political Studies and Public Administration
Bucharest
Romania

ISSN 2510-442X ISSN 2510-4438 (electronic)
Numanities - Arts and Humanities in Progress
ISBN 978-3-319-91985-0 ISBN 978-3-319-91986-7 (eBook)
https://doi.org/10.1007/978-3-319-91986-7

Library of Congress Control Number: 2018943707

Printed on acid-free paper

This Springer imprint is published by the registered company Springer International Publishing AG part of Springer Nature
The registered company address is: Gewerbestrasse 11, 6330 Cham, Switzerland

Contents

Chapter 1
Introduction

Alin Olteanu, Andrew Stables and Dumitru Borţun

This volume problematizes the relation between semiotics and communication studies, drawing on both theoretical arguments and matters of practical application. The idea behind this book has been the topic of discussion for a long time between the book's editors and the contributing authors. A key moment in this process consisted in the conference *Semiosis in Communication: Knowing and Learning* which took place in 2016 at the National University of Political Studies and Public Administration (Bucharest, Romania), organized in collaboration with Roehampton University (London, UK) and under the auspices of the International Association for Semiotic Studies. This conference brought together the editors of this volume and some leading researchers in these fields of research. At that moment the idea of this book started being drafted. The present papers were gathered with the aim of offering comprehensive and fresh insights on this interdisciplinarity.

Semiotics and communication theories are commonly studied together, in complementarity, while also considered to overlap. Meaning, the subject of semiotics, is arguably the most salient concept for communication. Also, the study of meaning has little purpose other than offering insights on communication. A tendency can be noticed in semiotics of regarding knowledge acquisition as a matter of communication. For instance, in explaining the basics of biosemiotics, the semiotic theory of biology, Thomas Sebeok starts by defining life forms as entities which communicate (1991: 9). Life is characterized by knowledge exchange, from

A. Olteanu (✉) · A. Stables
Kaunas University of Technology (KTU), Kaunas, Lithuania
e-mail: alin.olteanu@ktu.lt

A. Stables
e-mail: andrew.stables@icloud.com

D. Borţun
National University of Political Studies and Public Administration,
Bucharest, Romania
e-mail: dumitru.bortun@comunicare.ro

A. Olteanu et al. (eds.), *Meanings & Co.*, Numanities - Arts and Humanities
in Progress 6, https://doi.org/10.1007/978-3-319-91986-7_1

the endosemiosis of cells and organs within a metabolism, to the exosemiosis between organisms, inclusive of the most complex technological extensions of human means of expression. Social semiotics as well, finds its rationale in conjunction to communication theory (Kress and van Leeuwen 2001 , Kress 2010). Recently, both in its approach to biological organisms and to socio-cultural phenomena, semiotic theory tends to endorse a multimodal notion of meaning, dismissive of the classic double articulation hypothesis of linguistics, arguably originating in Saussure (1959/1916) and properly coined by Martinet (1962) (see Martinelli 2010: 91–93, Kress 2010: 1–11, Lotman 1994: 379). Semiotic approaches to communication also challenge the Sapir-Whorf hypothesis of linguistics, namely the idea that language parallels culture and acculturation and, consequently, shapes scientific research to various extents (Randviir and Cobley 2010).

Nevertheless, semiotics and communication theories also manifest some tensions. Arguably, on drawing on a rather technical and rigid notion of *information*, communication theories often ignore the complexity of meaning (see Fiske 1990: 39–40). Living organisms, obvious in the case of humans, have a wide range of semiotic competences to extract (discover or produce) meaning from various semiotic resources. As such, organisms are equipped with various degrees of freedom in interpreting data and, accordingly, in modeling their environments. As Peirce insisted (e.g., CP 4.153, 4.611, 6.302), there is always an element of chance in interpretation. This qualifies mechanistic approaches to communication as reductivist.

Thus, the present volume discusses this epistemological relation, its consequences for theory and suggests possibilities for future research. The papers gathered are addressed to a broad academic public, ranging from students to senior researchers, who have an interest in communication in general and, in particular, in semiotics and concepts of relevance for this relation, such as multimodality, intertextuality, code and modelling. Given the interdisciplinary and eclectic approaches and the interest for recent and innovative theories with a high potential for social impact, the volume rightly fits within the purpose of the Numanities (Martinelli 2016).

References

Fiske, John. 1990. *Introduction to Communication Studies*, 2nd ed. London Routledge.

Kress, Gunther. 2010. *Multimodality: A social semiotic approach to contemporary communication*. London: Routledge.

Kress, G., and T. van Leeuwen. 2001. *Multimodal discourse: The modes and media of contemporary communication*. London: Arnold.

Lotman, Yury. 1994. The text within the text. *PMLA* 109 (3): 377–384.

Martinelli, Dario. 2010. *A critical companion to Zoosemiotics: People, paths, ideas*. Dordrecht: Springer.

Martinelli, Dario. 2016. *Arts and humanities in progress: A manifesto of numanities*. Cham: Switzerland.

Martinet, André. 1962. *A functional view of language*. Oxford: Oxford University Press.
Peirce, Charles S. 1931–1935 and 1958. *The collected papers of Charles Sanders Peirce*, vols. I–VI, eds. Charles Hartshorne and Paul Weiss, vols. VII–VIII, ed. Arthur W. Burks. Cambridge MA: Harvard University Press.
Randviir, Anti, and Paul Cobley. 2010. Sociosemiotics. In *The Routledge companion to semiotics*, ed. P. Cobley, 118–134. London: Routledge.
Saussure, Ferdinand de. 1959/1916. *Course in general linguistics*, eds. Charles Bally and Albert Sechehaye, trans. Wade Baskin. New York, Toronto, London: McGraw-Hill Book Company.
Sebeok, Thomas. 1991. *A sign is just a sign: Advances in semiotics*. Bloomington and Indianapolis: Indiana University Press.

Part I
Theoretical Developments

Chapter 2
Modelling Human Communication: Mediality and Semiotics

Lars Elleström

Abstract The article delineates a model of communication among human minds that is designed to work equally well for all kinds of nonverbal and verbal significance. The model thus allows for detailed analysis of and comparison among all varieties of human communication. In particular, the transitional stage of communication, which is termed the media product, is thoroughly developed and conceptualized in terms of mediality and semiosis. By way of mapping basic media dissimilarities that are vital for differing communicative capacities, the model explains both how various sorts of significance can be transferred among minds, and why they cannot always be realized. The article furthermore suggests a semiotic framework for conceptualizing the familiar notion that communication is also strongly predisposed by surrounding factors such as earlier experiences and cultural influence.

2.1 Aim

The aim of this article is to delineate an irreducible model of human communication that is centered on the notion of medium, and furthermore to suggest how such an irreducible model may be connected to a semiotic conception of how cognitive import is formed in the mind of the perceiver of media products. The focus will thus be on the notion of medium. Medium means 'middle', 'interval', 'interspace', and so on; it is a word that can and should be used in a plenitude of very different circumstances. Here it will be delimited as a term to be used within the framework of communication among human minds. The notion of medium is thus to be understood as the intermediate stage of all kinds of inter-human communication; hence the term *medium* here refers to not only mass media, but also media used in more intimate communication; to both media based on external technological

L. Elleström (✉)
Linnaeus University, Växjö, Sweden
e-mail: lars.ellestrom@lnu.se

A. Olteanu et al. (eds.), *Meanings & Co.*, Numanities - Arts and Humanities in Progress 6, https://doi.org/10.1007/978-3-319-91986-7_2

devices, and media based on corporeality; to both premeditated media, and casual media; and to both media used for practical purposes, and artistic media.

Many successful and highly useful attempts to describe and analyze the basic features of human communication have already been made, and several of the most influential scholarly conceptualizations of the entities present within communication have taken the form of models (see Lanigan 2013). Nevertheless, the initial aim of this article is to expound a new model of communication (see also Elleström forthcoming a). Given the plethora of extant theories, why is yet another communication model called for?

First, a new model is necessary because, in reality, most existing models only take into account verbal significance while largely ignoring nonverbal significance or in effect subsuming it under the verbal—or at least under symbolic, conventionalized meaning-making. Whereas it is common to acknowledge that communication includes not only speech and other verbal media types, these admissions are often chiefly nominal and have had no far-reaching consequences for how communication has been theorized. I claim that it is deceptive to assume that the communication of nonverbal significance can be modelled by way of analogue-to-verbal significance.

I also argue that in order to remedy this lack we need a communication model that methodically identifies the irreducible components of communication, systematically puts them in relation to each other, and—most importantly—thoroughly develops the transitional stage of communication that is often called the 'channel', 'message', 'contact', or sometimes 'medium'. Although several existing models capture most of the essential communication entities in highly useful ways—classical models such as Hall (1980), Jakobson (1960), Schramm (1971), Shannon (1948), and several others that more or less follow their logics (see Elleström forthcoming a)—some core features of communication have yet to be properly incorporated. I maintain that these features are best conceptualized in terms of mediality and with the aid of semiotics, thereby offering theoretical tools that are vital for pinpointing basic media similarities and differences, and developing the material and mental aspects of mediality to a sufficient degree of complexity.

The task of a communication model is undoubtedly to offer, as lucidly as possible, a theoretical framework for describing, explaining, and analyzing processes of communication. While this must include an understanding of not only how communication is conceivable at all but why communication is often not possible to fully realize, there are vital sides of the latter aspect that have not yet been properly investigated. The technological notion of noise (originating from Shannon 1948) has offered only limited possibilities to explain communicative limitations beyond physical disturbances; imperative as the concept may be in engineering information data, it becomes rather trivial when applied to the communication of 'meaning'. In contrast, a well-developed notion of medium that includes both pre-semiotic and semiotic media traits offers the possibility to explain both how various sorts of significance can be communicated and why they cannot always be realized.

Therefore, basic media dissimilarities that are vital for differing communicative capacities must be mapped.

The term *medium* has certainly been applied before in communication models. In fact, probably most of the entities, qualities, and phenomena that will be highlighted in this article have been discussed in terms of media and mediation by various authors. To the best of my knowledge, however, the term has not yet been tied to a developed and systematic concept that takes into account all those fundamental qualities that are vital for a proper understanding of the intermediate stage of communication (see, for instance, the overview in Crowley 2013).

The article starts with a thorough delineation of the fundamental components of my proposed model of communication and then continues to discuss these components and their interrelations. After that, some vital conceptual clarification and distinctions are made before the central notion of a media product is investigated in detail. This is followed by some elaborations on the sometimes complex interrelations between communicating minds. Finally, a semiotic framework is suggested for conceptualizing the well-known notion that communication is also strongly influenced by surrounding factors such as earlier experiences and cultural influence.

2.2 The Irreducible Components of Communication

A model should be understood as a clearly outlined cognitive scheme that is both described with the aid of language and depicted as a diagram. My proposed model of inter-human communication consists of what I take to be the smallest and fewest possible entities and their essential interrelations. If one of these entities or interrelations is removed, communication is no longer possible; thus, the model is irreducible. I submit that three indispensable and interconnected entities can be discerned:

(1) Something being transferred;
(2) Two separate places between which the transfer occurs; and
(3) An intermediate stage that makes the transfer possible.

The first entity is often referred to as the *message* (Jakobson 1960; Schramm 1971). While the second entity arguably consists of two units, they can only be outlined in relation to each other. Whereas Jakobson's (1960) terms for these entities are *addresser* and *addressee*, Schramm (1971) prefers *communicator* and *receiver*. The third entity corresponds to Jakobson's "contact" that notably incorporates both a material and a mental aspect: "a physical channel and psychological connection between the addresser and the addressee" (1960: 353). Schramm used the term *message* to represent not only the transferred entity, but also the intermediate stage of communication (the message seems to be understood as something that is both 'transferred' and 'transferred through'). Importantly, however,

Schramm described the transmitting message not only as a material entity—such as "a letter"—but also as "a collection of signs", thus indicating the capacity of the material to produce mental significance through signs (1971: 15).

As Hall (1980) has indicated, transfer of meaning is very likely to entail change of meaning; this modification may be only slight or more radical. Nevertheless, I claim that communication models cannot do without the first entity, the notion of something being transferred. If there is no correlation at all between input and output, there is simply no communication, given the foundational idea that to communicate is to share; thus, a concept of communication without the notion of something being transferred is actually nonsensical. However problematic it may be, the notion of something being transferred must be retained and painstakingly scrutinized instead of being avoided.

My suggestion is to use the term *cognitive import* to refer to those mental configurations that are the input and output of communication. The notion that I want to suggest in using this term clearly is closely related to notions captured by terms such as *meaning* and *ideas*, although the term *cognitive import* is perhaps less burdened with certain notions that a term such as *meaning* seems to have difficulties getting rid of; meaning is often understood to be a rather rigid concept of verbal, firm, definable, or even logical sense. In contrast, cognitive import should not be confined to distinct or even consciously intended conceptions; it should rather be understood as a very broad notion of significance that is relevant to a wide range of media types. It is imperative to emphasise that although cognitive import is always a result of mind-work, it is not always possible to articulate using language; hence, communication, according to my proposed model, cannot be reduced to communication of verbal or verbalizable meaning.

The second entity (two separate places between which the transfer occurs) is most often construed as two persons. However, this straightforward notion is not precise enough for my purposes. As it is imperative to be able to connect mind and body to different entities of the communication model, one must avoid crude notions such as that of Jakobson's (1960) addresser-addressee and Schramm's (1971) communicator-receiver. These notions give the impression that the transfer necessarily occurs between two persons consisting of minds and bodies and with a third, separate, intermediate object in the middle, so to speak—an intermediate object in the form of a "message" that is essentially disconnected from the communicating persons. It is better to follow Hall's (1980) implicit idea that communication occurs between sites that are capable of holding "meaning". Weaver's (1998/1949) description of communication as something that occurs between "one mind" and "another" is right to the point.

My suggestion is to use the terms *producer's mind* and *perceiver's mind* to refer to the mental places in which cognitive import appears; first, there are certain mental configurations in the producer's mind, and then, following the communicative transfer, there are mental configurations in the perceiver's mind that are at least remotely similar to those in the producer's mind.

Most researchers referred to in this article have either explicitly or implicitly recognized that the third entity (an intermediate stage that makes the transfer possible) is in some way material. As stated succinctly in a recent publication, any act of communication “is made possible by some form of *concrete reification* of the message, which, at its most elementary level, must abide by physical laws to exist and take shape” (Bolchini and Lu 2013: 398). Furthermore, Schramm (1971) and Hall (1980) clearly discuss the intermediate stage in terms of signs. In line with this, I suggest that the intermediate entity connecting two minds with each other is always in some way material, although it clearly cannot be conceptualized only as materiality. Since it connects two minds in terms of a transfer of cognitive import, it must be understood as materiality that has the capacity to trigger certain mental responses.

My suggestion is to use the term *media product* to refer to the intermediate stage that enables the transfer of cognitive import from a producer’s to a perceiver’s mind. As the material bodies of these two minds may well be used as instruments for the transfer of cognitive import, they are potential mediators for media products. I propose that media products may be realized by either non-bodily or bodily matter (including matter emanating directly from a body), or a combination of these. This means that the producer’s mind may, for instance, use either non-bodily matter (i.e., the paper of a written letter) or her/his own body and its immediate extensions (i.e., forming speech and gestures) to realise media products. Furthermore, the perceiver’s body may be used to mediate media products; for instance, the producer may realise a painting on the perceiver’s skin or push her/him gently to communicate the desire that she/he move a bit. In contrast to influential scholars such as Marshall McLuhan who conceptualise media as the “extensions of man” in general (McLuhan 1994/1964), I thus define media products as ‘extensions of mind’ in the context of inter-human communication.

Since being a media product should be understood as a function rather than an essential property, virtually any material existence can be used as one, including not only solid objects but all kinds of physical phenomena that can be perceived by the human senses. In addition to those forms of media products that are more commonly categorized as such (written texts, songs, scientific diagrams, warning cries, road signs, etc.), there is an endless row of forms of physical objects, phenomena, and actions that can function as media products given that they are perceived in situations and surroundings that encourage interpretation in terms of communication. These include nudges, blinkings, coughs, meals, ceremonies, decorations, clothes, hair styles, and make-up. Also, dogs, wine bottles, and cars of certain breeds, brands, and fabricates may well function as media products to communicate the embracing of certain values or simply wealth, for instance. Within the framework of a trial, both the footage from surveillance cameras and spoken word from witnesses function as media products as well as fingerprints, DNA samples, and blood stains presented by the prosecutor.

Thus, there is no such thing as a media product ‘as such’. Not even a written text is a media product in itself, I argue; it is only when its function of transferring cognitive import among minds is realized that it can be conceptualized as a media product. The archaeologist who watches the marks on a bone and believes that they are caused by accidental scraping is not involved in communication. If the archaeologist believes that the marks are some sort of letters in an unknown language, the archaeologist may well be engaged in elementary communication to the extent that a communicative intent is understood. If the marks are eventually deciphered, more complex communication may result. If the deciphering actually turns out to be mistaken, the belief that communication occurred is an illusion. Border cases like these could of course be exemplified also by everyday interaction among people who may or may not be mistaken about the significance of all kinds of movements, glances, and sounds.

Given this conceptualization, it is pointless to try to distinguish between physical existences that are and that are not actual media products. What is important is rather to have a clear notion of the properties of physical existences that render them the function of media products. Clearly, these properties, which I will investigate in the following, are in no way self-evidently present. Perceiving something as a media product is a question of being attentive to certain kinds of phenomena in the world. As humans have been able to communicate with each other for thousands and thousands of years, this attention is partly passed on by heredity, but it is also deeply formed by cultural factors and the experience of navigating within one's present surroundings. Knowledge of musical performance traditions, for example, leads to specific attention to certain details while others may be ignored; thus, accidental noises and random gestures may be sifted out as irrelevant for the musical communication and not part of the media product. More practical knowledge of the environment normally makes us pay attention to what happens on the screen of a television set rather than to its back side. If the television set is used in an artistic installation, however, or if a repairman tries to explain why it does not work by way of pointing to certain gadgets, it may well be the back side that should be selected for attention in order to achieve the function of a media product. Media products must thus often be understood as cultural entities that are very much dependent on social praxis. They are formed by (often shared) selective attention on sensorially perceptible areas that are believed to be relevant for achieving communication.

2.3 A Medium-Centered Model of Communication

Before presenting the qualities that are vital for a proper understanding of the intermediate stage of communication—the media product—I will display my suggested communication model in the form of a visual diagram (Fig. 2.1) and elaborate on some of its implications. Construing this diagram from left to right, the act of communication starts with a certain cognitive import in the producer's mind. Consciously or unconsciously, the producer forms or somehow simply directs attention to a potential media product that may be taken in by some perceiver. The media product makes a transfer of cognitive import from the producer's mind to the perceiver's mind possible; this is a transfer not in the strong sense that the cognitive import as such passes through the media product (which lacks consciousness), but

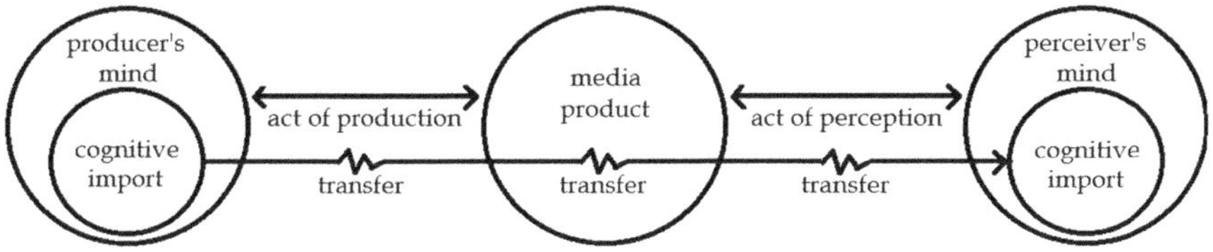

Fig. 2.1 A medium-centered model of communication

in the sense that there is, in the end, a cognitive import in the perceiver's mind that bears some resemblance to the cognitive import in the producer's mind.

The visual diagram contains the three entities of communication circumscribed above:

(1) Something being transferred: cognitive import;
(2) Two separate places between which the transfer occurs: producer's mind and perceiver's mind; and
(3) An intermediate stage that makes the transfer possible: media product.

Additionally, it displays four essential interrelations among these entities:

(1) An act of production 'between' the producer's mind and media product;
(2) An act of perception 'between' the media product and the perceiver's mind;
(3) Cognitive import 'inside' the producer's mind and the perceiver's mind; and
(4) A transfer of cognitive import 'through' the media product.

I will now elaborate on these interrelations, especially the fourth one. The notion of a media product, which was only preliminarily demarcated above, and the question of how cognitive import may be transferred through a media product are essential for any attempt to understand the core of communication.

The first interrelation (an act of production 'between' the producer's mind and media product) is always initiated by the producer's mind, and always, to begin with, effectuated by the producer's body. Sometimes this primary bodily act immediately results in a media product, for instance, when one person begins talking to another person who is standing beside her: the speech emanating from the vocal chords constitutes a possible media product that reaches the perceiver directly. At other times the primary bodily act is linked to subsequent stages of production, and it is not unusual for the primary bodily act to be connected to a broad range of actions and procedures before a media product comes to be present for a perceiver. For instance, talking through a telephone often requires manual handling of the telephone in addition to the activation of the user's vocal chords, and always requires constructed, technological devices that are suitable to transmit the initial speech to another place in which the actual media product is constituted—that is, the speech that can be heard and understood by the perceiver. Similarly, a child drawing a picture for her father who is sitting at the same kitchen table only has to perform, in principle, one primary bodily act in order to create a media product that is immediately available for the perceiver. However, if the father is in another place, additional stages of actions and procedures must be supplemented: the drawing may be posted and physically relocated, or scanned and emailed, after which it appears in a slightly transformed way as a media product that is mediated by a computer screen. The act of production may thus be simple and direct as well as complex and indirect. It may furthermore include stages of storage.

The second interrelation (an act of perception 'between' the media product and the perceiver's mind) is always initiated by the perceiver's sense organs and always, to some extent, followed by and entangled with interpretation. Interpretation should be understood as all kinds of mental activities that somehow make sense of the sensory input; these activities may be both conscious and unconscious, and are no doubt already present in a basic way when the sense impressions are initially processed. Thus, compared to the potentially extensive act of production, the act of perception is brief and very quickly channeled into interpretation, which of course occurs in the perceiver's mind. Nevertheless, the type, quality, and form of sensory input provided by the media product, and actually taken in by the perceiver's sense organs, are absolutely crucial for the interpretation formed by the perceiver's mind.

A comparison of 'meaning'-oriented communication models, such as the one that I propose here, and Shannon's (1948) famous model of transmission of quantifiable data, may give the impression that the acts of production and perception are equivalent to the entities and processes that Shannon calls transmitter/signal and received signal/receiver. However, the production and perception of media products involve complex and entangled corporeal and cognitive processes that cannot properly be reduced to analogies of mechanical processes. Therefore, Weaver's (1998/1949) suggested extension of Shannon's model to the area of cognition and signification only works on a rather superficial level. Whereas in an engineering context it may well be true that "The *receiver* is a sort of inverse transmitter, changing the transmitted signal back into a message" (Weaver 1998/1949: 7), the act of perception cannot accurately be described as an inverse act of production unless one is satisfied with a very rudimentary analogy. The creation of cognitive import requires much more than an unbiased perception of the raw material qualities of the media product; it is not a process of reproduction of computable data, but rather one of interpreting perceived sensory configurations—it involves semiosis. Although this problem is not entirely neglected by Weaver, he does not scrutinise it at any length. Furthermore, leaving the claims of Weaver aside, it must be noted that the act of production often actually involves the producer's acts of perception and interpretation of the emerging media product that brings the two acts closer to each other. Rather than saying that the act of perception is an inversion of the act of production, which makes the relation between the two acts resemble a purely material transmission, one must say that the act of production and the act of perception are related in an act of material and mental transmission, including transition.

Even if one notes that there is a shallow analogy between Shannon's (1948) model of communication of computable data and 'meaning'-oriented communication models, it is more useful to observe that Shannon's (1948) model might actually be incorporated as part of the act of production in my proposed model. Whereas some acts of production do not require external technological devices, others do. As the elementary examples of person-to-person speech and telephone calls demonstrated, it may be the case that a primary bodily act immediately results in a media product, while at other times the primary bodily act is linked to subsequent stages of production involving external technology that ultimately result in a media product that is presented to the perceiver. Needless to say, Shannon's (1948) model and mathematical theory has proven exceptionally useful for the engineering of the non-corporeal stages of production, while it is largely irrelevant for the corporeal and mental stages of production. The "destination" in Shannon's (1948) model of communication can thus not be understood as a person, as claimed; an act of perception must be added if his model is to be applicable to the communication of cognitive import. Consequently, the solution is to understand the entities of Shannon's (1948) model as potential parts of the act of production in my suggested model, with "destination" seen not as a person but as a physical existence —the media product—that must be perceived and interpreted by the perceiver's mind if communication of cognitive import is to be achieved.

The third interrelation among the entities of communication (cognitive import 'inside' the producer's mind and the perceiver's mind) will only be briefly commented upon here. Clearly, one cannot state without intricate implications that there is a certain amount of confinable cognitive import inside a mind, and it is undoubtedly difficult to judge the actual extent of similarity between the two amounts of confined cognitive import in the two minds. Deciding this in a more precise way is probably beyond the reach of presently known research methods. However, I find the notion unproblematic that the transferred cognitive import is only one part of the producer's and the perceiver's minds; the cognitive import is 'inside' the minds in the respect that it is closely interconnected with a multitude of other cognitive entities and processes, and in the end with the total sum of mental activities in general that surrounds it. I will come back to the fundamental importance of the surrounding areas of the mind in the last sections of the article.

The fourth interrelation (a transfer of cognitive import 'through' the media product) is central for my arguments in this article. So far, the media product has mainly been described as the entity of communication that enables a transfer of cognitive import from a producer's mind to a perceiver's mind; a material existence that has the capacity of triggering a mental response. In order to give a somewhat more detailed account of this notion, the very capacity itself must be scrutinized.

Of course, the transfer of cognitive import is only partly comparable to other transfers such as, for instance, the transfer of goods between two cities by train. The cognitive import transfer is not a material transfer, but a mental transfer with the aid of materiality. In one respect it can be compared to teleportation, the transfer of energy or matter between two points without traversing the intermediate space: the cognitive import is indeed transferred between two points (two minds), and contrary to the transfer of goods, it does not traverse the intermediate space. Nevertheless, as the transfer depends on the media product, it may reasonably be said to go 'through' it. The media product is actually neither a neutral object of material transfer, like a freight car, nor an intermediate space without effect, as in teleportation; it constitutes a crucial stage of transition in addition to transmission.

2.4 Media Products and Technical Media

At this stage of the account, it is necessary to introduce a delicate but sometimes vital distinction between media products and what may be called *technical media*, or to be more precise, *technical media of distribution of sensory configurations* or *technical media of display*. I define a technical medium as any object, physical phenomenon, or body that *mediates*, in the sense that it 'realizes' and 'displays' the entities that we construe as media products [cf. Müller's distinction between "technical conditions" and "media products" (1996: 23; see also 81–82)]. I have emphasized that media products are material existences fulfilling a communicative function; however, media products need technical media in order to come into existence. As media products, technical media are defined by their function; although there are many objects that are manufactured to work as technical media, any physical existence may acquire the function of mediating a media product. Technical media of display should thus be understood as entities that have the capacity to present media products and make them available for the senses of the perceiver; they distribute sensory configurations (Elleström 2010: 30–37, 2014b: 47–56). For instance, Innis (1950) emphasizes the importance of technical media such as stone, clay, papyrus, and paper for the historical development of communication—more specifically writing—and society at large. More modern technical media include electronic screens and sound waves produced by loudspeakers. All these different kinds of physical entities are necessary conditions for making media products discernible.

As I define the notion of technical medium, it is narrower than the notion of "physical media", for instance, as circumscribed by Clüver (2007: 30). Devices used for the realization of media products, but not tools used only for the production or storage of media products, are technical media. The brush and the typewriter are tools of production that are normally separated from the material manifestations of media products and are as such not technical media according to my definition, although they count as physical media in Clüver's (2007) sense. If used in the ways that they are primarily designed for, the flute and the video camera are partly tools for production and partly devices for the realization of media products and can hence also be seen as technical media. Some physical existences, such as ink on paper, may both store and display media products and thus work as technical media, whereas a computer hard disk—a device for storage only—is not normally a technical medium in the sense that is emphasized here.

Technical media of display are thus physical substances, devices, or phenomena that make it possible for media products to be physically manifest. They clearly exist in very diverse forms. Above, I suggested that media products may be realized by either non-bodily or bodily matter. In other words, there are external technical media (non-bodily materialities such as clay, screens, ink on paper, sound waves from loudspeakers, or just about anything chosen from the surroundings) and there are internal technical media (bodies, parts of bodies, or physical phenomena emanating directly from bodies, such as a voice). All forms of external and internal technical media may be combined with each other in countless ways. They may either simply be at hand in the environment of the producer's mind (like an empty glass used to communicate the desire to get something to drink by way of directing the waiter's attention to it), they may be more or less crafted (like a piece of paper used for displaying the words 'one more beer, please'), or they may be corporeal actions (like a movement of hand and arm imitating the act of drinking). These examples do not in any way exhaust the many possible modes of existence for technical media, but I will not further dwell on this matter.

The distinction between media product and technical medium is clearly theoretical rather than being a distinction between two different kinds of material entities. On the contrary, the technical medium is a prerequisite for the physical existence of a media product, and in a communicative situation the perceiver identifies only one level of presence: the perceived sensory configurations emanating from some physical existence. The distinction is needed in order to demonstrate the difference—and mutual interdependence—between a piece of music (a media product) and a music audio system (a technical medium) displaying it, for example.

Some technical media, such as audio systems, are well fitted for being reused many times. This is also the case for a technical medium such as a television set (which actually consists of two kinds of technical media: a flat screen that emits photons and loudspeakers that set the air into pulsation) that may realise several different media products (many television programs). A communicating human body may be conceptualized in a similar fashion: when it is moved in certain ways and in certain circumstances, it mediates gestures. As long as the memory of these gestures is stored in the brain, similar gestures can be performed by the same technical medium of display—the body—thus creating a very large amount of equivalent media products. The body may of course also be used for mediating billions of very different media products.

On the other hand, some technical media tend to be used only once or a few times. A marble block being cut to a certain form mediates a sculpture, and it can normally be reused only a limited number of times. As the block not only displays but also stores the sculptures, the reuse of the technical medium implies the destruction of the initial media product.

It must be noted, though, that common language far from always provides words for properly describing the distinction between technical media of display and media products. This is because of the boundless and dynamic nature of human communication: for reasons of mental economy, only the most common and salient media products are categorized and given names. One may exemplify with the communicative acts performed by the thirsty person that I recently discussed. The movement of hand and arm is used as a technical medium for mediating what is commonly known as a gesture, a kind of media product. The paper is used as a technical medium for mediating a media product that may be called, for instance, a written note. The raised empty glass, however, makes some resistance to being described in ordinary language; one may say that 'glass' or 'a glass' is used as a technical medium, but what kind of a media product does it mediate? This is not crystal clear. Nevertheless, the media product is there with or without a proper word to denote it.

Finally, here are some notes on the duplication of media products. The notion of a media product, according to my definition, implies that every single display through a technical medium constitutes a specific media product. This display may last for a very short time (a cry of warning, for instance), for a very long time (such as a rock-painting), or anything in between. In any case, the display of such media products may be repeated in various ways. Several cries of warning are being heard, and several rock-painting are being seen, and some of these are very similar. In some cases, the similarity between media products is so detailed that it is more than reasonable to think that they are 'the same'. When I again watch the movie *Fantasia,* I certainly believe that it is the same movie that I saw some years ago, having the same title and being identical in virtually all details, although it was then displayed on the screen in a movie theatre and not on the screen of my television set.

Yet, the two mediations of *Fantasia* are not the same media product. On a theoretical level, it is important to be able to acknowledge that every single mediation of a media product is unique, although several media products may be extremely similar indeed (like the thousands of copies of operating instructions for a certain kind of toaster). On a pragmatic level, however, it is efficient to operate with the notion of sameness. The life outside the domain of scholarly articles would become very difficult to handle if we did not recognise that different persons, at different times, and located at different places, may actually watch 'the same television program', such as a specific episode of *Monty Python's Flying Circus*, although they perceive different media products that are generally virtually undistinguishable but slightly different when it comes to qualities such as the size and resolution of the moving images and the quality of the sound. Under theoretical pressure, however, the 'sameness' of different actual mediations becomes diffuse and problematic. Are my operating instructions, full of coffee stains and almost illegible, the same media product as your unstained operating instructions? As they can hardly communicate the same cognitive import (understanding how to handle the toaster), I would say not. If one still argues that two unstained operating instructions are the same, the obscure question arises: how many stains or torn pages are required to render them different? In the end, the question of sameness becomes a somewhat metaphysical question. Therefore, strictly speaking, different media products may only be the same in the respect that they are very similar. Although different media products are never ontologically the same, they may be thought of as being 'the same' in many other important respects. Perhaps one could say that very similar media products are variations of an abstract but recognizable communicational composition that may be reproduced more or less efficiently.

2.5 Media Modalities

Having explained the intricate relationship between technical media and media products, it is now time to map the common features of media products in some detail. I suggest that all media products, without exception, can be analyzed in terms of four kinds of basic traits that may be called *media modalities* (Elleström 2010). This calls for some explanation. As postulated earlier, the media product is the entity though which cognitive import is transferred in communication. It requires a technical medium of display to be realized. The defining qualities of technical media of display, I argue, are their capacities to realise specific kinds of sensible interfaces between media products and perceivers' minds. The perception of the media product is furthermore deeply entangled with cognitive operations that may broadly be called interpretation.

As this complex process of the transfer of cognitive import involves both material and mental aspects, I find it helpful to distinguish between two profoundly interrelated but nevertheless discernible basic facets of the communicative process: *mediation* and *representation*. Mediation is the display of sensory configurations that are perceived by human sense receptors within a communicative situation. It is a pre-semiotic phenomenon and should be understood as the physical realization of entities (with material, sensorial, and spatiotemporal qualities, and semiotic potential). For instance, one may hear some sound. Representation is a semiotic phenomenon and should be understood as the core of signification, which is delimited in this research to how humans create cognitive import in communication. When a human agent forms sense of the mediated sensory configurations, sign functions are activated and representation is at work. For instance, one may interpret the heard sound as a voice uttering meaningful words.

My current emphasis is on the notion that both a pre-semiotic and a semiotic side exist to basic encounters with media. Whereas the concept of mediation highlights the material realization of the medium, the concept of representation highlights the semiotic conception of the medium. Although mediation and representation are clearly entangled in complex ways, upholding a theoretical distinction between them is vital. This theoretical distinction is helpful in analyzing complex relations and processes. However, in practice, mediation and representation are deeply interrelated. Every representation is based on the distinctiveness of a specific mediation. Furthermore, some types of mediation facilitate certain types of representation and render other types of representation impossible. As an obvious case in point, vibrating air emerging from the vocal chords and lips that is perceived as sound but not as words is well suited for the iconic representation of bird song, whereas such sounds cannot possibly form a detailed, three-dimensional iconic representation of a cathedral. However, distinctive differences among mediations are frequently subtler and less easily spotted without close and systematic examination.

We can now return to the notion of media products having four kinds of basic traits called media modalities. Three of these modalities are pre-semiotic, which means that they cover media traits that are involved in signification—the creation of cognitive import in the perceiver's mind—although they are not semiotic qualities in themselves. The three modalities are thus not *a*-semiotic; they are *pre*-semiotic, meaning that the traits that they cover are bound to become part of the semiotic as soon as communication is established. The pre-semiotic traits concern the fundamentals of mediation, which is to say that they are necessary conditions for any media product to be realized in the outer world by a technical medium, and hence for any communication to be brought about.

The three pre-semiotic media modalities are the material modality, the spatiotemporal modality, and the sensorial modality. Media products are all material in the plain sense that they may be, for instance, solid or non-solid, or organic or inorganic, and comparable traits like these belong to the material modality. It is also the case that all media products have spatiotemporal traits, which means that such products that do not have at least either spatial or temporal extension are inconceivable; hence, the spatiotemporal modality consists of comparable media traits such as temporality, stasis, two-dimensional spatiality, and three-dimensional spatiality. Furthermore, media products must reach the mind through at least one sense; hence, sensory perception is the common denominator of the media traits belonging to the sensorial modality—media products may be visual, auditory, tactile, and so forth.

Of course, these kinds of traits are not unknown to communication researchers. For instance, Hall (1980) discusses the two sensory channels of television; Berlo (1960) highlights all five external senses; and Schramm (1971) at least briefly mentions that "a message has dimensions in time or space" (32). However, thorough understanding of the conditions for mediation requires systematic attention to all three pre-semiotic modalities. It is clear that cognitive import of any sort cannot be freely mediated by any kind of material, spatiotemporal, and sensorial traits. For instance—to make some blatant examples—complex assertions cannot easily be transferred through the sense of smell, and it is more difficult to effectively transfer a detailed series of visual events though a static media product than through a temporal media product.

The fourth media modality is the semiotic modality that covers media traits concerning representation rather than mediation. Whereas the semiotic traits of a media product are less palpable than pre-semiotic ones, and in fact are entirely derived from them (because different kinds of mediation have different kinds of semiotic potential), they are equally essential for realizing communication. The mediated sensory configurations of a media product do not transfer any cognitive import until the perceiver's mind comprehends them as signs. In other words, the sensations are meaningless until they are understood to represent something through unconscious or conscious interpretation. This is to say that all objects and phenomena that act as media products have semiotic traits by definition.

As discussed in this research, representation is the creation of cognitive import through the perceptual and cognitive acts of reception. I submit that to say that a media product represents something is to say that it triggers a certain type of interpretation. This interpretation may be more or less hardwired in the media product and the manner in which one perceives it with one's senses, but it never exists independently of the cognitive activity of the recipient. When something represents, it calls forth something else; the representing entity makes something else—the represented—present in the mind. This is to say, in terms of Charles Sanders Peirce's foundational notions, that a sign or *representamen* stands for an *object*; Peirce's third sign constituent, the *interpretant*, may be understood as the mental result of the representamen–object relation [see for instance, 1932: CP 2.228–229 (c. 1897)]. My notion of cognitive import created in the perceiver's mind in communication is clearly an instance of Peirce's notion of interpretant. However, the whole triad of sign constituents is actually part of a mental process, although both representamens and objects may be (or rather may be connected to) external material elements or phenomena (see Elleström 2014a).

Whereas representation—the very essence of the semiotic—constantly occurs in our minds when we think without having to be prompted by sensory perceptions, it is also triggered by external stimuli; in this context, focusing on external stimuli resulting from mediation is appropriate. In other words, representation also occurs in pure thinking and in the perception of things and phenomena that are not part of mediation, but the account of representation in this research is limited to the creation of cognitive import on the basis of mediated sensory configurations—stimuli picked up by our sense receptors in communicative situations. The contention is that all media products represent in various ways as soon as sense is attributed to them. Hence, the media product can be understood as an assemblage of representamens that, due to their qualities (material, spatiotemporal, and sensorial traits), represent certain objects (that are available to the perceiver), thus creating interpretants (cognitive import) in the perceiver's mind. According to my terminology, the notion of non-representative media products is self-contradictory.

By far the most successful effort to define the basic ways to create sense in terms of signs is Peirce's foundational trichotomy icon, index, and symbol. These three basic sign types are defined on the basis of the representamen-object relationship and can be understood as fundamental cognitive abilities. Icons stand for (represent) their objects based on similarity; indexes do so based on contiguity; and symbols rely on habits or conventions (1932, CP2.247–249 (c.1903); Elleström 2014a: 98–113). I take iconicity, indexicality, and symbolicity to be the main media traits within the semiotic modality, which is to say that no communication occurs unless cognitive import is created through at least one of the three sign types (icons, indexes, and symbols).

Again, semiotics is certainly not unknown in communication research. Among the scholars quoted in this article, Schramm clearly related to some basic semiotic features. For instance, he accurately noted that "it is just as meaningful to say that B [the receiver] acts on the signs [the message], as that they act on B" (1971: 22). Indeed, the mind of the perceiver is very active in construing the signs of the media product. In addition, Hall spoke in terms of semiotics, although with a distinct linguistic bias. On the whole, I judge that semiotic approaches to communication based on the tradition of Ferdinand de Saussure that ignore or downplay the role of iconicity and indexicality have been harmful to the development of theory that also embraces non-verbal communication. Peirce's semiotic framework is much more fruitful as it incorporates sign types that work both within and far outside of the linguistic domain.

Furthermore, my emphasis here is on the notion that a semiotic perspective must be combined with a pre-semiotic perspective. Communication is equally dependent on the pre-semiotic media modalities and the semiotic modality. What we take to be represented objects called forth by representamens or signs (objects such as persons, things, events, actions, feelings, ideas, desires, conditions, and narratives), are results of both the basic features of the media product as such (the mediated material, spatiotemporal, and sensorial traits) and of cognitive activity (resulting in representation). While signification is ultimately about mind-work, in the case of communication this mind-work is fundamentally dependent on the physical appearance of the media product—although some representation is clearly more closely tied to the appearance of the medium, whereas other representation is more a result of interpretation, and hence the setting of the perceiving mind.

As with pre-semiotic traits, the semiotic traits of a media product offer certain possibilities and set some restrictions. Obviously, cognitive import of any sort cannot be freely created on the basis of just any sign type. For instance, the iconic signs of music can represent complex feelings and motional structures that are probably largely inaccessible to the symbolic signs of written text; conversely, written symbolic signs can represent arguments and the appearance of visual objects, with much greater accuracy compared to auditory icons. Flagrant examples like these are only the tip of the iceberg in terms of the (in)capacities of signs based on similarity, contiguity, and habits or conventions, respectively. Therefore, communicative transfer of cognitive import through a media product is made possible—but also profoundly limited—by the semiotic traits of the medium.

In line with this proposal, it is appropriate to bring the earlier briefly mentioned notion of noise back into the discussion. Shannon's (1948) idea that signal disturbances in communication can be conceptualized as noise has been picked up by many researchers engaged in the communication of 'meaning'. The basic phenomenon of disruptions that occur on the way from the producer's to the perceiver's mind is clearly relevant to the transfer of cognitive import. For instance, speech can be disturbed by other sounds and a motion picture can be disrupted because of material decay or censorship. Noise in this sense occurs in both the act of production and the act of perception. In my visual model of communication (Fig. 2.1), this noise is shown as disruptions in the arrow representing transfer of cognitive import—both before and after the transfer through the media product—reflecting possibly unsatisfactory conditions of production and perception.

The problem with the notion of noise when applied to the communication of cognitive import is that it might imply that complete absence of noise would bring about complete transfer of cognitive import—as in the case of technical transmission of computable data—which is clearly not the case. The technological notion of noise is simply not sufficient to understand the communication of cognitive import. According to Hall, "distortions" or "misunderstandings" are also due to, among other things, "the asymmetry between the codes of 'source' and 'receiver' at the moment of transformation into and out of the discursive form" (1980: 131).

This is definitely a step in the right direction in terms of offering a more complex notion of possible disruptions in the communication of cognitive import. However, it does not provide a more complete view of restraining factors in the transfer of cognitive import. It must also be emphasized that creators of media products generally do not have access to—or do not master—more than a few media types. Consequently, they often cannot possibly form media products that have the capacity to create cognitive import in the perceiver's mind that is similar to the cognitive import in their own mind. Therefore, I argue that perhaps the most fundamental restraining factors of communication are to be found in the basic pre-semiotic and semiotic traits of the media products.

Many exceedingly complex factors are clearly involved when the perceiver's mind forms cognitive import. My proposed model highlights one cluster of crucial factors in particular: media products have partly similar and partly dissimilar material, spatiotemporal, sensorial, and even semiotic traits, and the combination of traits to a large extent—although certainly not completely—determines what kinds of cognitive import can be transferred from the producer's mind to the perceiver's mind. Songs, emails, photographs, gestures, films, and advertisements differ in various ways concerning their pre-semiotic and semiotic traits, and hence can only transfer the same sort of cognitive import to a limited extent. In my diagram (Fig. 2.1), this communicative restriction is shown as disruptions in the arrow representing the transfer of cognitive import as it passes through the media product.

2.6 Producer's and Perceiver's Minds

Outlining only the smallest and fewest possible entities of communication and their essential interrelations, my suggested model of communication (Fig. 2.1) is irreducible but certainly expandable. In the preceding Sect. 2.5 I have already fleshed it out through suggesting various ways of conceptualizing the notion of media product in some detail. I will now also sketch out a more multifaceted comprehension of the minds of the producer and the perceiver and their interrelations.

The minimal level of complexity consists of simply one mind producing a single media product that is perceived by another mind. This, I believe, is the essential core of human communication. In actual communicative situations, however, the perceiver's mind is very often also a producer's mind. On the basis of the cognitive import generated by an initial media product, the perceiver becomes a producer in creating another media product (of the same or another kind) that reaches an additional perceiver's mind, thereby forming new cognitive import that is more or less similar to that in earlier producers' minds. Hence, a communicative chain is formed. When the communicative chain involves the initial producer and perceiver constantly changing roles and forming new media products (of the same or another kind), we have two-way communication. The creation of new media products in two-way communication is often conceptualized as feedback that may result in the creation of cognitive import that is either rather constant or significantly developed. Needless to say, communicative chains that are uni- and bidirectional may be combined in a multitude of ways.

It is furthermore often the case that media products are produced or perceived by several minds. For instance, a motion picture is normally both produced and perceived by more than one mind. While the minds of scriptwriters, directors, actors, and many more together create the motion picture, the audience consists of a multitude of perceiving minds. In contrast, a plenary talk is as a rule produced by one mind but perceived by many. An unsuccessful theatre performance may be produced by many minds but perceived (from an off-stage position) by only one.

Another level of complexity consists of the case when perceivers take in their own media product. Although I would not say that pure thinking is communication [as suggested by Berlo (1960: 31)], perception of one's own media product created earlier may mean that the mind tries to construe cognitive import on the basis of the media product rather than on the memory of what one had in mind on the occasion of production. In this case, a transfer of cognitive import actually occurs through a media product from one mind to another, in the sense that the mind, when perceiving the media product, is in a different state compared to that during production. The effort of writing a scholarly text is a good example of this sort of internal communication: alas, sometimes communication fails when one cannot understand one's own words written the day before.

Of course, this level of complexity can also be combined with others, as in the case of interactive video games. Such games are normally constructed and designed by several minds, but the point here is that the actual media products (the many realized sensory configurations that are mediated by screens and sounding loudspeakers each time the game is being played) are also created by the players. We thus have a kind of communication involving several producing minds that have created certain frames for interaction and resulting consequences (when designing the game), one or several producing minds that create the actual media product in their interaction with the evolving media product (when playing the game), and one or several perceiving minds that are actually the same as those minds that interact with and hence produce the media product: the specific realization of the possibilities of the video game. Naturally, this media product may also be perceived by additional minds that are not co-producers (i.e., an audience).

It thus turns out that the irreducible notion of the producer's mind and perceiver's mind may well be simple but certainly not reductive. On the contrary, it offers a solid basis for analyzing all kinds of communicative complexities. While the examples above do not exhaust the intricacies, they may hint at the many complicated ways in which producers' and perceivers' minds may be positioned in various communicative circumstances.

2.7 Intra-communicational and Extra-Communicational Domains

In addition to developing the basic notion of transfer of cognitive import between *two separate minds*, I will now also elaborate on the notion of cognitive import *in* the producer's and especially the perceiver's mind. As the irreducible model of communication only states that cognitive import is transferred between minds, it is appropriate to suggest not only a way of understanding how it is formed by basic media traits (which was done in the section on media modalities), but also a way of comprehending how it is molded by surrounding factors. In addition to its innate basic capacity to perceive and interpret mediated qualities, the mind is inclined to form cognitive import on the basis of acquired knowledge, experiences, beliefs, expectations, preferences, and values—preconceptions that are largely shaped by individual experiences of culture, society, geography, and history. It is clear that this concept is immensely important for the outcome of communication. The perceiver's mind acts upon the perceived media product on the basis of both its hardwired cognitive capacities and its attained predispositions; evidently, the cognitive import that was stored in the mind before the media product was perceived has a significant effect—to various degrees—on the new cognitive import formed by communication.

This is a widely recognized fact that has been extensively theorized in various ways; I can only mention here a few researchers who have already been referred to in this article. Jakobson (1960) discusses it in terms of "a context [that is] seizable by the addressee, and either verbal or capable of being verbalized" (353). Context is no doubt important for all kinds of communication, although I think it is a mistake —even for a restricted focus on verbal communication—to say that the context must be able to be verbalized in order to be relevant. Hall (1980) distinctly emphasizes the "social relations of the communication process as a whole" and the "frameworks of knowledge" (129–130), and discusses them in detail. These and other issues that are central to the formation of cognitive import in a broad context have been minutely scrutinized within the research area of hermeneutics. I will here suggest a complementary semiotic way of circumscribing how cognitive import in communication is formed by surrounding factors. Although focus is on the perceiver's mind, the basic suggested principles are also relevant for the formation of cognitive import in the producer's mind.

We have already established that the representamens that initiate semiosis in communication come from sensory perception of media products. One perceives configurations of sound, vision, touch, and so forth that are created or brought out by someone and understood to signify something; they make objects (in the Peircean sense) present to one's mind—and eventually result in interpretants based on the representamen-object relation: it is these interpretants that constitute the cognitive import being transferred in communication. Yet, from where do the objects come? They clearly do not emerge out of nothing, but are drawn forth from earlier percepts, sensations, and notions that are stored in the perceiver's mind, either in long-term or short-term memory that may also cover ongoing communication. 'Earlier' may hence be a century ago or a fraction of a second ago.

In semiotic terms, the stored mental entities may be direct percepts from outside of communication, interpretants from semiosis outside of communication, interpretants from semiosis in earlier communication, or interpretants from semiosis in ongoing communication. This is to say that objects of semiosis always require "collateral experience" (Peirce 1958: CP 8.177–85 (1909); cf. Bergman 2009) that may derive both from within and without ongoing communication. In other words, collateral experience may both be formed by semiosis inside the spatiotemporal frame of the communicative act and stem from other, earlier involvements with the world, including former communication as well as direct experience of the surrounding existence.

In line with this twofold origin of collateral experience, I distinguish between two utterly entwined, but nevertheless dissimilar areas in the mind of the perceiver of media products: the *intra-communicational* and the *extra-communicational domains*, emphasizing a difference between the forming of cognitive import in ongoing communication and what precedes and surrounds it (Elleström forthcoming b; related but divergent distinctions in cognitive psychology have been proposed by Brewer 1987: 187). I also find it appropriate to make a corresponding distinction between *intra-communicational* and *extra-communicational objects*, both formed by collateral experience from their respective domains.

From a broad temporal perspective, the extra-communicational domain is clearly prior to each new intra-communicational domain created. That is why I will roughly delineate the extra-communicational domain before exploring the intra-communicational domain. However, it will eventually become clear that the theoretical framework sketched here actually has its center in the intra-communicational domain; my conceptual constructions are molded from the point of view of this domain. The discernment of something extra-communicational is obviously a consequence of focusing on what is intra-communicational.

The extra-communicational domain should thus be understood as the background area in the mind of the perceiver of media products. It comprises everything one is already familiar with. As it is a mental domain, it does not consist of the world as such, but rather of what one believes and knows through perception and semiosis. This is to say that one's stored experiences not only consist of percepts as such, but also of percepts that have been contemplated and processed by the mind through semiosis. This involves estimations and evaluations of encounters with people, societies, and cultures that are consciously or unconsciously accepted, put in doubt, or rejected. The extra-communicational domain includes experiences of both what one presumes to be more objective states of affairs (dogs, universities, music, and statistical relations), what one presumes to be more subjective states of affairs (states of mind related to individual experiences), and everything in between. It is thus actually formed in one's mind not only through semiosis and immediate external perception, but also through interoception, proprioception, and mental introspection. The extra-communicational/intra-communicational domain distinction is hence very different from exterior/interior to the mind, world/individual, material/mental, and objective/subjective.

It is important to note that vital parts of the extra-communicational domain are constituted by perception and interpretation of media products. Former communication is thus very much part of what precedes and surrounds ongoing communication. Together, non-communicative and communicative prior experiences form "a horizon of possibilities", to borrow an expression from Ryan (1984: 127); the extra-communicational domain is the reservoir from which entities are selected to form new constellations of objects in the intra-communicational domain.

In contrast to the extra-communicational domain, the intra-communicational domain is the foreground area in the mind of the perceiver of media products. It is formed by one's perception and interpretation of the media products that are present in the ongoing act of communication. It is based on both extra-communicational objects—emanating from the extra-communicational domain—and intra-communicational objects—arising in the intra-communicational domain—that together result in interpretants making up a salient cognitive import in the perceiver's mind. However, the intra-communicational domain is largely mapped upon the extra-communicational domain. Rehashing Ryan's "principle of minimal departure" (1980: 406), I argue that one construes the intra-communicational domain as being the closest possible to the extra-communicational domain, and allows for deviations only when they cannot be avoided. In other words: one does not question familiar ideas and experiences until it is called for.

As the intra-communicational domain is formed by communicative semiosis, it may be called a *virtual sphere*. The virtual should not be understood in opposition to the actual but as something having potential. I hence define the virtual as a mental sphere, created by communicative semiosis, that *has the potential* to have *real connections* to the extra-communicational. This is to say that a virtual sphere may possibly represent extra-communicational objects indexically; it may be truthful to the extra-communicational domain (this notion is further elaborated in Elleström forthcoming b).

A virtual sphere can consist of many kinds of cognitive import: anything from a brief thought triggered by a few spoken words, a gesture, or a quick glance at an advertisement, to a complex narrative or a scientific theory formed by hours of watching television or reading books. Depending on the degree of attention to the media products, the borders of a virtual sphere need not necessarily be clearly defined. As communication is generally anything but flawless, a virtual sphere may be exceedingly incomplete or even fragmentary. It may also comprise what one apprehends as clashing ideas or inconsistent notions. As virtual spheres consist of cognitive import resulting from communication, they are (per definition) to a certain extent shareable among minds.

The coexistence of intra-communicational and extra-communicational objects results in a possible double view on virtual spheres: from one point of view, they form self-ruled spheres with a certain degree of experienced autonomy; from another point of view, they are always exceedingly dependent on the extra-communicational domain. The crucial point is that intra-communicational objects cannot be created ex nihilo; in effect, they are completely derived from extra-communicational objects. This is because nothing can actually be grasped in communication without the resource of extra-communicational objects. Even the most fanciful narratives require recognizable objects in order to make sense (cf. Bergman 2009: 261). To be more precise: intra-communicational objects are always in some way parts, combinations, or blends of extra-communicational objects; or, to be even more exact: intra-communicational objects are parts, combinations, or blends of interpretants resulting from representation of extra-communicational objects.

It is possible to represent griffins, for example, (which, to the best of our knowledge, exist only in virtual spheres) because of one's acquaintance with extra-communicational material objects such as lions and eagles that can easily be combined. A virtual sphere may even include notions such as a round square, consisting of two mutually exclusive extra-communicational objects together forming an odd intra-communicational object. Literary characters such as Lily Briscoe in Virginia Woolf's novel *To the Lighthouse* are composite intra-communicational objects consisting of extra-communicational material and mental objects that stem from the world as one knows it. One cannot imagine Lily Briscoe unless one is fairly familiar with notions such as walking, talking, and eating; what it means to refer to persons with certain names; what women and men, adults and children are; what it means to love and to be bored; and what artistic creation is. Also, more purely mental extra-communicational objects may be modified or united into new mental intra-communicational objects. Objects such as familiar emotions may be combined into novel intra-communicational objects consisting of conflicts between or blends of emotions that are perceived as unique, although one is already acquainted with the components.

The question then arises: if all intra-communicational objects are ultimately derived from extra-communicational objects, how come virtual spheres are often experienced as having a certain degree of autonomy? This is because they, in part or in whole, may be perceived as new *gestalts* that disrupt the connection to the extra-communicational domain. This happens when one does not immediately *recognise* the new composites of extra-communicational objects. The reason for their not being re-cognized is that they have not earlier been cognized in the particular constellation or merger in which they appear in the virtual sphere. Several such disruptions lead to greater perceived intra-communicational domain autonomy. Even though intra-communicational objects are entirely dependent on extra-communicational objects, one may thus say that they emerge within the intra-communicational domain.

The relation between extra-communicational and intra-communicational objects may be even more complex than hitherto indicated. Intra-communicational objects that are perceived as new gestalts might, in their turn, be part of more embracing gestalts that *are* recognized from the extra-communicational domain. A virtual sphere can contain representations of intra-communicational objects such as living trains formed by familiar extra-communicational objects such as 'man-made machines for transportation' and 'the quality of being animate and conscious' that together form a new gestalt. When living trains quarrel or fall in love with each other, however, they interact in a way that is directly recognized from the extra-communicational domain. The conclusion is that intra-communicational objects may be interspersed among extra-communicational objects in numerous, complicated ways.

2.8 Virtual Sphere, Other Virtual Spheres, Perceived Actual Sphere

Having described the interrelations between the intra-communicational and the extra-communicational domains in some detail, I will now present an overview with the aid of a visual diagram (Fig. 2.2). The main point is that whereas the intra-communicational domain simply consists of one virtual sphere, the extra-communicational domain consists of two rather different elements: on the one hand, other virtual spheres; on the other hand, what I propose to call the *perceived actual sphere*. From the point of view of a virtual sphere, there are thus three more or less distinct spheres: the virtual sphere itself, other virtual spheres, and the perceived actual sphere.

The *perceived actual sphere* consists of *extra-communicational, immediate, and presented* material and mental objects *beyond the realm of communication* that the perceiving mind is acquainted with. 'Perceived' shall be understood in a broad

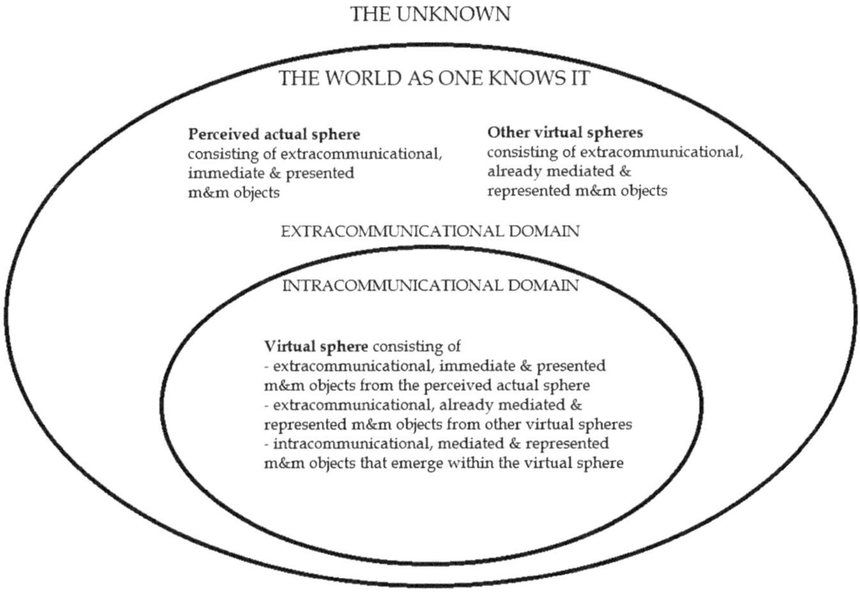

Fig. 2.2 Virtual sphere, other virtual spheres, perceived actual sphere

sense to include exteroception, interoception, and proprioception, and also to be joined with mental introspection and semiosis based on perception of the actual sphere. 'Immediate and presented' shall be understood in contrast to communication: the perceived actual sphere does not consist of mediated representations formed by media products brought out by minds and their extensions, but is *immediately present* to us. Note that immediately present does *not* mean without the mediating mental mechanisms that connect sensation to perception, or the complicated mediating functions that connect perception to the external world.

The *other virtual spheres* consist of *extra-communicational, already mediated, and represented* material and mental objects that the perceiving mind is acquainted with. As virtual spheres are thoroughly semiotic, these objects are always made of former interpretants. The other virtual spheres result from communication and comprise mediated representations formed by media products brought out by minds and their extensions.

Hence, the *virtual sphere* consists of *extra-communicational, immediate and presented* material and mental objects from the perceived actual sphere + *extra-communicational, already mediated and represented* material and mental objects from other virtual spheres + *intra-communicational, mediated and represented* material and mental objects *that emerge within the virtual sphere.*

As all schematic representations, this outline is intended to give an overview of an intricate reality. It nevertheless not only points to mental areas that are fundamentally different in certain respects, but also reveals their complex interrelations. It must thus be emphasized that every virtual sphere, from the point of view of that sphere, is intra-communicational, and hence composed of objects that are derived from itself (to the extent that parts, combinations, and blends of extra-communicational objects may be understood as distinct), as well as from other virtual spheres and the perceived actual sphere. This comprises a *mise-en-abyme*: intra-communicational virtual spheres are formed by perceived actual spheres and by other, extra-communicational virtual spheres that are in turn formed by perceived actual spheres and by other, extra-communicational virtual spheres ad infinitum.

Adding the diagram in Fig. 2.2 to the diagram in Fig. 2.1 might give a sense of how the irreducible model of communication may be expanded in terms of a more complex understanding of the transferred cognitive import. In brief, the totality of the intra-communicational and extra-communicational domains in Fig. 2.2 (the outer circle) is equivalent to the whole perceiver's mind in Fig. 2.1 (the outer circle); the intra-communicational domain, comprising the virtual sphere in Fig. 2.2 (the inner circle), consists of the cognitive import in the perceiver's mind according to Fig. 2.1 (the inner circle). This virtual sphere is not only formed by the perception and interpretation of the specific traits of the media products that are present in the ongoing act of communication, as emphasized in Fig. 2.1; it is simultaneously based on both extra-communicational and intra-communicational objects that together result in interpretants making up a salient cognitive import in the perceiver's mind, as demonstrated in Fig. 2.2. In other words, the cognitive import in the perceiver's mind, bringing about a virtual sphere, is formed by both the ongoing experience of the particular traits of the media product and the general collateral experiences of all sorts in the perceiver's mind.

The extent to which cognitive import may be shared among the producer's and perceiver's mind is thus partly determined by how much the extra-communicational domain of the perceiver's mind overlaps with the extra-communicational domain of the producer's mind (understood as the background area in the mind of the producer of media products). This conclusion corresponds well with established views on the importance of shared experiences and values for successful communication.

2.9 Conclusions

The irreducible model of communication presented in this article, with its clearly defined entities and interrelations among entities, has several advantages. It avoids the widespread but notoriously indistinct notion of a message that generally conflates communicative entities. As a model of the transfer of cognitive import from one mind to another through a media product, it allows for a refined pinpointing of the minds and bodies of the communicators in separate parts of the model; this does not rule out the vital fact that mind and body are profoundly interrelated. Most importantly, it is not molded upon a verbal template, but allows for detailed analysis of and comparison among all kinds of communication.

The heart of the proposed model consists of the media product. Although the true complexity of this intermediate entity has only been hinted at, some rudimentary clarifications have been made. A methodical, bottom-up conception of material, spatiotemporal, sensorial, and semiotic media traits paves the way for avoiding blunt and often misleading dichotomies such as 'text' versus 'image', or 'verbal' versus 'visual' media. Instead, basic media similarities and dissimilarities can be analyzed in a more refined way that makes it possible to more sharply discern the obstacles of transferring cognitive import among minds.

As displayed, the irreducible model is a good starting point for systematic elaboration of its entities and their essential interrelations, thus allowing for more complex conceptions derived from its ostensible plainness. As demonstrated in the final sections of the article, it also works in harmony with the recognized notion of communication being strongly influenced by surrounding factors related to the communicating individuals and the culture and society that they are situated in.

References

Bergman, Mats. 2009. Experience, purpose, and the value of vagueness: On C. S. Peirce's contribution to the philosophy of communication. *Communication Theory* 19: 248–277.

Berlo, David K. 1960. *The process of communication: An introduction to theory and practice*. New York: Holt, Rinehart, and Winston.

Bolchini, Davide, and Amy Shirong Lu. 2013. Channel. In *Theories and models of communication*, ed. Paul Cobley and Peter J. Schulz, 397–410. Berlin, Boston: De Gruyter.

Brewer, William F. 1987. Schemas versus mental models in human memory. In *Modelling cognition*, ed. Peter Morris, 187–197. Oxford: Oxford University Press.

Clüver, Claus. 2007. Intermediality and interarts studies. In *Changing borders: Contemporary positions in intermediality*, ed. Jens Arvidson, Mikael Askander, Jørgen Bruhn, and Heidrun Führer, 19–37. Lund: Intermedia Studies Press.

Crowley, David. 2013. Mediation theory. In *Theories and models of communication*, ed. Paul Cobley and Peter J. Schulz, 309–325. Berlin, Boston: De Gruyter.

Elleström, Lars. 2010. The modalities of media: A model for understanding intermedial relations. In *Media borders, multimodality and intermediality*, ed. Lars Elleström, 11–48. Basingstoke: Palgrave Macmillan.

Elleström, Lars. 2014a. Material and mental representation: Peirce adapted to the study of media and arts. *The American Journal of Semiotics* 30: 83–138.

Elleström, Lars. 2014b. *Media transformation: The transfer of media characteristics among media*. Basingstoke: Palgrave Macmillan.

Elleström, Lars. Forthcoming a. A medium-centered model of communication. *Semiotica*.

Elleström, Lars. Forthcoming b. Coherence and truthfulness in communication: Intra-communicational and extra-communicational indexicality. *Semiotica*.

Hall, Stuart. 1980. Encoding/decoding. In *Culture, media, language: Working papers in cultural studies, 1972–79*, 128–138. London: Hutchinson.

Innis, Harold A. 1950. *Empire and communications*. Oxford: Oxford University Press.

Jakobson, Roman. 1960. Closing statement: Linguistics and poetics. In *Style in language*, ed. Thomas Sebeok, 350–377. New York: Wiley.

Lanigan, Richard L. 2013. Information theories. In *Theories and models of communication*, ed. Paul Cobley and Peter J. Schulz, 59–83. Berlin and Boston: De Gruyter.

McLuhan, Marshall. 1994/1964. *Understanding media: The extensions of man*. Cambridge MA, London: MIT Press.

Müller, J.E. 1996. *Intermedialität. Formen moderner kultureller Kommunikation*. Münster: Nodus Publikationen.

Peirce, Charles Sanders. 1932. *The collected papers of Charles Sanders Peirce*, vol. 2, ed. Charles Hartshorne and Paul Weiss. Cambridge MA: Harvard University Press.

Peirce, Charles Sanders. 1958. *The collected papers of Charles Sanders Peirce*, vol. 8, ed. Arthur W. Burks. Cambridge MA: Harvard University Press.

Ryan, Marie-Laure. 1980. Fiction, non-factuals, and the principle of minimal departure. *Poetics* 9: 403–422.

Ryan, Marie-Laure. 1984. Fiction as a logical, ontological, and illocutionary issue. *Style* 18: 121–139.

Schramm, Wilbur. 1971. The nature of communication between humans. In *The process and effects of mass communication*, revised ed, ed. Wilbur Schramm and Donald F. Roberts, 3–53. Urbana: University of Illinois Press.

Shannon, Claude E. 1948. A mathematical theory of communication. *The Bell System Technical Journal* 27 (379–423): 623–656.

Weaver, Warren. 1998/1949. Recent contributions to the mathematical theory of communication. In *The mathematical theory of communication*, 1–28. Urbana and Chicago: University of Illinois Press.

Chapter 3
Exploring a Semiotic Conceptualisation of Modelling in Digital Humanities Practices

Arianna Ciula and Cristina Marras

Abstract Digital Humanities (hereafter DH) is a research field engaged in exploring how humanities scholarship is transformed and extended by the digital and vice versa. The core practice of DH research is modelling which implies the translation of complex systems of knowledge into computationally processable models. In our work we contextualise DH practices within a semiotic framework; namely we consider modelling as a strategy to make sense (signification) via practical thinking (creation and manipulation of models). A semiotic approach of this kind contributes to stress the dynamic nature of models and modelling, and to reinstate in renewed terms the understanding of modelling as open process of signification enacting a triadic cooperation (among object, *representamen* and interpretant). Referring to Peirce classification of hypoicons, we reflect on some DH examples of modelling in the form of images, diagrams and metaphors, claiming that a semiotic understanding of modelling could ultimately allow us to surpass the duality object versus model (as well as sign vs. context). We thus propose to consider modelling as a creative and highly pragmatic process of thinking and reasoning in which metaphors assume a central role and where meaning is negotiated through the creation and manipulation of external representations combined with an imaginative use of formal and informal languages.

A. Ciula (✉)
King's Digital Lab, King's College London, London, UK
e-mail: arianna.ciula@kcl.ac.uk

C. Marras (✉)
Institute for European Intellectual Lexicon and History of Ideas,
Italian National Research Council, Rome, Italy
e-mail: cristina.marras@cnr.it

A. Olteanu et al. (eds.), *Meanings & Co.*, Numanities - Arts and Humanities in Progress 6, https://doi.org/10.1007/978-3-319-91986-7_3

3.1 Introduction

Digital Humanities is a research field engaged in exploring how humanities scholarship is transformed and extended by the digital and vice versa. This mutual transformation and extension concern tools as well as epistemologies. The core practice of DH research is modelling (McCarty 2005: 20–72; Buzzetti 2002; Flanders and Jannidis 2015), which implies the translation of complex systems of knowledge or conceptual frameworks into computationally processable models or operational frameworks.

In our work[1] we contextualise DH practices within a semiotic conceptualisation of modelling (Ciula and Marras 2016; Ciula and Eide 2014, 2017) adapted from Kralemann and Lattmann (2013) and complemented by intermedia theories on iconicity (Elleström 2013). A semiotic framework allows us to see modelling primarily as a strategy to make sense (signification) via practical thinking (creating and manipulating models). In fact, a semiotic approach gives high prominence to a dynamic view on models and modelling which reinstates in renewed terms the understanding of modelling as open process, that is a process of signification enacting a triadic cooperation (amongst object, *representamen* and interpretant).

By reflecting on some DH examples of modelling (classified as images, diagrams and metaphors), we will claim that a semiotic understanding of modelling could ultimately allow us to surpass the duality object versus model (as well as sign vs. context) and see modelling as a creative process of thinking and reasoning where meaning is negotiated through the creation and manipulation of external representations combined with an imaginative use of formal and informal languages.

The paper is divided into four parts: the underpinning concepts behind our ongoing research project (see note 2) are introduced to set the floor for the argument which follows (Sect. 3.2); DH and semiotics are related dialectically to discuss three central concepts, namely iconicity, modelling, and translation (Sect. 3.3); finally, we explore the potentialities of an integrated use of languages (metaphorical and literal) in DH modelling paving the way for an interdisciplinary and pragmatic approach (Sect. 3.4) to modelling.

[1]The work for this essay was conducted as part of the research project *Modelling between digital and humanities: thinking in practice* (http://www.modellingdh.eu). Project partners: Arianna Ciula, University of Roehampton (until January 2017) and King's Digital Lab (King's College London, from February 2017); Øyvind Eide, University of Cologne; Cristina Marras, Institute for European Intellectual Lexicon and History of Ideas, National Research Council (Rome); Patrick Sahle, University of Cologne. The project is funded from April 2016 to July 2018 under the Volkswagen Stiftung programme: *"Original—isn't it?" New Options for the Humanities and Cultural Studies*, Funding Line 2 "Constellations" (2016–2017).

3.2 Modelling Between Digital and Humanities: Thinking in Practice

The project *Modelling between digital and humanities: thinking in practice* has an interdisciplinary context and departure aiming to:

(1) Explore the possibilities for an interdisciplinary language around modelling spanning humanities, cultural studies and sciences;
(2) Study modelling from a Digital Humanities perspective, that is to say focusing on modelling as anchored both to critical humanities and to the craft of engineering and computer sciences.

The underpinning idea is to be able to develop connections between modelling as both research and learning strategies ("learning by doing") but also to grasp the everyday engagement with modelling as a form of modern cultural literacy. Indeed, models are becoming ubiquitous; they are used to schematise the complex world we live in. Unpacking the rhetoric of (data) models is becoming an essential 'reading' skill or critical competence to master.

While it is important not to get stuck into convoluted analytical definitions, we need a pragmatic stance on what doing Digital Humanities means to contextualise our work further. Starting from an overview on the binomial 'Digital Humanities', Table 3.1 shows the different semantic aspects and functions (facets) being evoked by the use of 'digital' associated with areas or objects of scholarship.

The last row is where DH, as we intend it here, sits. As stated in the prospectus of the Design and Technology MA programme at Srishti Institute of Art (Bangalore, India):

> […] digital humanities explores how the questions posed in humanities scholarship are transformed and extended by the digital – both by means of tools and epistemologies (Ray Murray: http://srishti.ac.in/programs/pg-program-ma-in-digital-humanities).

Indeed, some of us in Digital Humanities might lean towards a preferred interest on tools rather than epistemologies (how we come to know) or vice versa, but

Table 3.1 Terminology and facets (adapted from Ciula 2017b)

Terminology	Facets
Digital e.g. Digital versus analogue	Representational e.g. form of data
Computational e.g. computational linguistics	Instrumental e.g. formal manipulation of symbols; process-ability of data
Digital + e.g. digital humanities	Methodological (epistemological) e.g. how data are created and understood; augmenting, extending humanistic methods with science, technology, engineering, mathematics and vice versa

ultimately the intersection between the two is where the most fruitful prospects for the field lie (Ciula 2017a). Recalling the use of the transformative plus evoked by Cecire (2011) and McCarty (2014) in combining 'Digital + Humanities', we can add that the transformation and expansion mentioned above can be mutual.

Modelling seems to be a useful and fruitful practice to revisit the relation between digital and humanities, for example, by forcing us to unpack the rhetoric of (data) models. There are however both terminological and operational challenges to overcome.

Indeed, the complexity of the concept of modelling is not only the result of decades of theory and practices in modelling, but it is deeply embedded in the roots and history of the term itself. *Model* and *to model* (modelling) belong to the same lexical family and are polysemic. Model, on the one hand, is used to identify formal or rule-based ways of reasoning and of representation (in mathematical and deductive modes of reasoning, for example); on the other hand, the term has also been related to the alternative and complementary processes of reasoning involving metaphorical, inductive, and abductive approaches. The former modes of reasoning associated to modelling have been well explored and theorized by scholars and they represent the framework of reference in the sciences (and arguably also in DH), while the latter approaches have not yet been 'fully recognized' as a form of modelling adequate to DH, nor have they been investigated as such.[2]

Therefore, it is not surprising that there are multiple and mostly discipline-dependent definitions of modelling. The practice of modelling in DH is mainly theorised around understandings of modelling adopted in the techno-sciences and computer science in particular (McCarty 2005, 2009; Flanders and Jannidis 2015; Mahr 2009). Recently, model-making was theorised within a semiotic framework (i.e. Kralemann and Lattmann 2013; Ciula and Eide 2014, 2017; Ciula and Marras 2016), whereby modelling is framed as a process of signification (semiotic process—meaning making).

Taking stock of both approaches and their intersections, by modelling we intend mainly two things (Ciula and Eide 2017: i33):

(1) A creative process of thinking and reasoning, where meaning is made and negotiated through creation and manipulation of external representations;
(2) A research strategy intended as a process by which researchers make and manipulate external representations ("imaginary concreta", Godfrey-Smith 2009) to make sense of objects and phenomena.

In line with Nersessian's (2008) *continuum* hypothesis adopted in her cognitive-historical account of modelling, (2) is nothing but a specification of (1), constrained and enhanced by the idiosyncratic contexts and purposes of research endeavours.

[2]The following discussion on terminology benefited from the research conducted by Michela Tardella within the project *Modelling between digital and humanities: thinking in practice*.

It should be stressed that the process of negotiation of meaning as outlined above involves both semiotic and terminological levels; it is strictly related to some epistemological questions concerning the nature of the objects being modelled (when we encounter the *model of* an existing object) or 'created' (when we encounter a *model for* an object to be) and more precisely the relation between modeller, modelling and model. If we assume that in a humanities research setting, the "Modelling process is part of what is being modelled" (Flanders 2012), it becomes crucial to ask what model of modelling can be "adequate" to DH.

It is precisely to address this question that we reflect below on three constitutive elements of humanities research: objects of study, practices of modelling, and languages of scholarship (including modelling languages).

3.2.1 Factuality and Fictionality of Objects

Rooted in the history of science and philosophy, the discussion concerning the nature of an object of experience has been led by crucial epistemological and philosophical questions: how can objects pertaining to the physical world (causally determined) have an effect on the human mind (generally considered as a free entity)? Under which conditions can the existence of external objects be processed?

Within the history of science, the problem is framed as a struggle to reach an accurate representation of nature and to find an adequate language able to provide the link between external objects, belief and knowledge (Daston 2000). Encompassing discussions spanning from the language of God to the 'language of nature' (Galileo Galilei); natural philosophers have traditionally addressed the problem by questioning the truth and falsity (in *senso lato*) of the objects of scientific knowledge and of the mathematical entities explaining them. Especially in the early modern era, the making and use of microscopes, telescopes and lenses changed the perspective, the distances and the accessibility to the smallest as well as the furthest objects. A process of approximation based on analogy identifying relational abstractions, mapping the structure of the object, defining, analysing and measuring, rather than 'directly' seeing objects, was introduced. Consequently the perception of the nature of the object changed radically from what could be directly measured and touched to something that is somehow 'created'. In early modern times the object of experience assumed a complex profile and was often conceived as the result of a cultural process and activity.

A still common and affirmed image of science today is that of an activity based on the experimental gathering of facts, a mathematical modelling of its results, and the framing of hypotheses and theories. This process, involving in particular the development and constitution of scientific objects, stresses the intrinsic interweaving of perception, representation and consciousness; it has its counterpart also in scholarly debates in the humanities, ranging from questions around the nature of consciousness (Palmieri 2012) and the nature of cognition, to the definition of the ethnographic object, just to mention a few.

Semiotics has contributed to this discussion with its theory of signs whereby a semiotic object is generally intended as to what a sign represents (or encodes) within a generative interpretative process. Socio and ethnographic disciplines within and outside the humanities consider cultural artefacts—spanning, for example, from weapons to inscriptions, from folk dances to theatre productions—as their objects of analysis. In the last century, material culture approaches have evolved across those disciplines to inform object-based analysis aiming at unpacking artefacts, write their biographies, their stories shaped by use ("thick descriptions", Geertz 1973, 1993). Within this perspective artefacts are considered intentional, cultural releasers "animated by their passage through the lives of people" (Graves-Brown 2000). While the total meaning of an artefact might be unattainable, the claim is that there are clues to the ways of thinking and living of those who made it. Similarly, in the realm of textual scholarship, theories connected to material culture approaches such as a social theory of texts recognise texts as open objects to be understood in the dynamic condition of their production and use.[3]

With respect to a DH research context, the privileged objects of modelling activities are, or at least have been so far, texts. In line with a social theory approach, by texts we intend here dynamic cultural objects (material documents as well as conceptual objects) contingent on the contexts of their production and reading, expressed in a wide range of manifestations from linear to discontinuous narrative, from manuscripts to printed editions, encompassing hybrid modalities.

Taking into consideration the evolution around the understanding of the nature of objects of experience outlined above, the objects that take part in an act of modelling—whether they are for example the texts being studied or the models being created—feature both an element of factuality (an experienced substance) and one of fictionality (they presuppose some rules of artifice). This implies that in a DH modelling activity a process of making explicit both components and their interaction is paramount; this means making explicit the perspectives of study towards the analysis of an object both in traditional and computational terms; so, for example, unpacking a theory of texts is a way of making explicit the fictionality of a modelling act as important as algorithm criticism.

3.2.2 Modelling Practice

If modelling acts are used to make sense of our cultural objects (mainly but not only texts), they are meaning-making practices and hence in themselves objects of study

[3]Jerome McGann has articulated this framework extensively over the years and made it relevant also to a DH research context; see his recent book (McGann 2014).

for the humanities.[4] Hence almost inevitably in DH, the modelling process is part of what is being modelled as outlined above.

Modelling is considered to be a or *the* core research methodology in Digital Humanities (McCarty 2005). In this sense it is intended as DH-specific research and teaching activity mainly embedded within a techno-scientific approach, but also connected to multifaceted conceptualizations of modelling as used and understood by other disciplines and practices.

In very sketchy and simplified terms, as modellers we behold a certain understanding of a cultural phenomenon of some sort. Via often informal external meta-models of aspects of such understanding (e.g. graphical representations like diagrams), formal and computable models of components of that understanding are generated and manipulated in an iterative cycle, which progressively—often via repeated and adjusted trials and errors—make us gain a more nuanced analysis of reduced portions of that phenomenon we departed from, sometimes even in conflict with our original takes on that phenomenon These iterative experimental cycles have been theorised extensively within industrial design practices engaged with making things. In connection with digital modelling in literary criticism, Moretti (2013) talks about a process of 'operationalization' whereby concepts of humanistic inquiry (e.g. Hegelian *pathos* in Greek tragedies) are literally operationalized, made measurable, hence creating empirical objects of study which bear theoretical consequences for the discipline of literary study as a whole:

> [Measurement] makes some concepts 'actual' in the strong sense of the word; it takes character-space, and proves that there is something in the real world (the real world of fictions) that corresponds to it. Not all concepts are born equal, some are better than others, and operationalization, though not the only test of a theory, is an important one. It shows that, by following a series of steps, you can turn abstractions into a clear and, hopefully, unexpected elaboration of reality. […] Operationalizing has certainly changed, and radicalized, our relationship to concepts: it has raised our expectations, by turning concepts into magic spells that can call into being a whole world of empirical data; and it has sharpened our skepticism, because, if the data revolt against their creator, then the concept is really in trouble. A theory-driven, data-rich research programme has become imaginable, bent on testing, and, when needed, falsifying the received knowledge of literary study. Of this enterprise, operationalizing will be the central ingredient. (Moretti 2013: 107–119)

What is interesting to note is that even in empirical very technical or implementation-focused settings, formal (e.g. computationally rule-bound) and informal (e.g. narrative, story-based) models coexist and interact to give sense to our modelling efforts.[5]

---

[4]This is of course based on a wide definition of what the objects of study for the humanities at large are:

The humanities study the meaning-making practices of human culture, past and present, focusing on interpretation and critical evaluation, primarily in terms of the individual response and with an ineliminable element of subjectivity (Small 2013: 57).

[5]For an in depth discussion of the role of formal models and stories in economics see Morgan and Knuuttila (2012).

In addition, Moretti's words make clear how the reconceptualization of a field theory can occur via modelling or what he calls operationalization, the computational methods themselves are subject to criticism, to refining and re-building, as part of one or more modelling cycles. As another example, in digital textual editing, one might adopt the Ordered Hierarchy of Content Objects (OHCO) model combined with a Text Encoding Initiative (TEI) document abstract model and schema; however, during the modelling process itself, the OHCO model as well as the TEI categorisations might be questioned deeply, for example because of their inadequateness to represent the modeller's theory of the historical documents at hand or because of the constraints the models force on the specific objects of interest:

> [The modeller] will most likely go back and forth in her modelling efforts to match – based on her knowledge and scholarly language – what she would like to elicit in the document (e.g. the organisation of the diplomatic formulas, the occurrence of names of witnesses in certain locations in the document, the abbreviations occurring in the date clauses) with the formal hierarchical structure proposed by one or more of the TEI guidelines chapters on the encoding of primary sources. Once this mapping reaches a certain stability and she is able to actually process that model, her interpretative activity will likely have to zoom in and out of these manifestations (e.g. between single cases and emerging patterns). She will often puzzle over exceptions that will allow her to iteratively refine the original mapping or require a total rethinking of categories and of the OCHO [sic] model being adopted. (Ciula and Marras 2016)

To conclude this section, we can state that modelling is a pragmatic activity framed within the complex cognitive, social, and cultural functioning of DH practices affected by cross-linguistic and interdisciplinary dimensions. As further discussed in Sect. 3.3 and elsewhere (Ciula and Marras 2016), a pragmatic stance highlights how the relational and dynamic aspects of modelling operate. Pragmatic modelling is conceived as being anchored to theory and language, while at the same time claiming some freedom from both.

3.2.3 Metaphorical Language

Language as mediation in designing and contextualising models is crucial given its regulative function in fixing and creating boundaries around the knowledge domain of reference. By observing the use of language in modelling activities, it emerges that the act of modelling and in particular its representative and descriptive functions operate within a context which includes a metaphorical language and which metaphors adapt to and also transform. In a sense, one might say that, however rich, the lexical resources of a language are sometimes unsatisfactory in capturing the totality of the speaker's expressive needs. The use of semantic means, such as the proliferation of lexemes or their polysemic use, barely increases the linguistic ability to satisfy these needs. Metaphors and other figures of speech become, in this respect, indispensable means to nurture the 'creative' direction of the use of language without expanding the semantic system in a strict sense. Metaphors involve both linguistic expressions and conceptual mappings (Marras 2017, 2013); they describe novelties, facilitate the understanding and interpretation of theories. This 'expansive' and heuristic force makes the metaphorical language particularly suitable to and fruitful in modelling activities and in science more in general. Recognising this force contributes to overcome the distinction between a rhetorical use of metaphors and a scientific 'cognitive' one.

Metaphors and analogies are central to Western scientific thought. They have been largely used by scientists in building the foundations of experimental and theoretical sciences, in representing and describing a domain, and in promoting understanding of scientific investigations (Marras 2006). Indeed, as recognised for decades now, a metaphor, given its cognitive and creative resonance, is much more than an episodic linguistic phenomenon. In addition, basic conceptual metaphors are never actually *cashed out* in a non-metaphorical language; in other words it is rare that basic conceptual metaphors shaping also our scientific language and rooted in the tradition of thought are 'retranslated' into their literary meaning. Conceptual metaphors have become a common vehicle to understand scholarly theories. For example, machines with their (directly observable) interacting parts and causal operations are used as the most powerful model for explaining natural events and became one of the most used conceptual metaphors in the history of science and philosophy (Haken et al. 1993).

DH are not immune to metaphorical language (even in the overarching discourse about the definition of the discipline as in as in McCarty 2006; Marras 2014). A metaphorical lexicon pervades the DH discourse, where technical terminology and metaphors are intertwined (e.g. surfing, navigating, artificial intelligence, computer machinery, etc.).[6] In DH metaphors are also used for structuring concepts at a macro-level, as for example in the use of 'ecosystem' or the cartography and landscape metaphors used to organise contents and visualise knowledge. Nevertheless, there is something specific related to the process and language of modelling in DH that we would like to stress here. Indeed, a metaphorical language is not only present in the DH discourse as stressed above but it has a heuristic function. In the data analysis, acquisition and storage DH applies established and mainstream techniques (such as for example textual markup in XML, relational models for structuring databases), along with more refined presentation and visualisation methods (charts, graphs, interactive maps, network analysis, timelines). In both cases the language describing and conceptualising the process of modelling has to take into account and grasp the interpretative activity at stake (what are the data? how are they created and understood?) as well as the dynamics of the relations between data/objects being analysed/mapped and the model/conceptualization resulting from it. Metaphorical expressiveness constitutes in our view an irreplaceable part of the linguistic lexicon of modelling. The key claim here is that metaphors assume a complex role and function tentatively spelled out as following:

---

[6]For a discussion on the presence of metaphors in DH complemented by some examples related to research projects see Ciula and Marras (2016).

- Metaphors themselves are models of knowledge (or of a specific conceptual space under study).[7] They define what knowledge is, what are the approaches to and visions of knowledge being adopted, what are the schemes within which knowledge or a specific concept operates. In short, they generate the cognitive and operational path to follow. For example, in a DH context, we can identify metaphors that conceptualise a broader disciplinary approach and guide data collection or acquisition as exemplified in Sect. 3.3.
- Metaphors enable the grasping of the unknown via what is known, both by making implicit knowledge explicit and by leveraging on unexpected connections with other semantic domains (see below). For example, in a DH context, the process of abstraction usually defined as analysis translates the idiosyncratic data of interest into thematic clusters and categories which in turn become useful to ingest and classify new data.
- Metaphors define their semantic fields, which have the potential to expand across related domains. In a DH context these affordances have practical implications, for example with respect to how data is processed and stored, and is organised in designing data models and data presentations or interfaces; very often the labelling of entities and their relations makes use of the semantic fields opened by the guiding modelling metaphor (e.g. in network models).
- Metaphors structure meaning and senses. In a DH context, this is where a metaphor translates into a data-model and data-structure (also at the level of presentation of the data). So for example, a hierarchical ontological model might define a specific idea of relation across entities, which implies the inheritance of properties.
- Metaphors lead narratives and reshape their contexts of production and use. In a DH context, the metaphorical language around a project influences the presentation of the data, the design of interfaces, and of course also the narratives in more traditional publication formats.

To sum up, the notion of metaphor we adopt is productive in understanding acts of modelling and is not restrictive; it does not consist on a purely verbal expression, nor does it corresponds simply to the use of a metaphor in an analogy-based model. Instead, the notion of metaphor we make use of conflates different aspects: cognitive, conceptual, heuristic, relational, and metalinguistic; indeed, in modelling as elsewhere, metaphors work transversally at the experimental, theoretical and practical level.

[7]The point of departure for the discussion on scientific models as metaphors in contemporary literature can be traced back to Max Black's interaction view of metaphor (1962) and Mary Hesse's seminal work *Models and Analogies in Science* (1966). For a discussion on metaphor and thought see the contributions collected in Ortony (ed.) (1993).

3.3 DH and Semiotics

In this section we would like to draw the attention to some key 'conceptual' intersections between DH and semiotics and to how, in some cases, semiotics concepts applied and used in a DH context acquire a specific 'meaning'. As mentioned above, at least two key cross concepts emerge as particularly relevant, namely iconicity and translation.

3.3.1 Iconicity and Its Grades

The discussion on iconicity and modes of reasoning is one of the leading themes in semiotics since its beginnings: 'it is by icons only that we really reason' (Peirce 1933, CP 4.127 [1893]).

The general semiotic meaning of iconicity is that of resemblance or analogy between the form of a sign (representamen, signifier, source) and its object (signified). What is interesting in DH, and in particular in the discussion around modelling, is not as much the redefinition of the concepts *vis à vis* the new disciplinary context but the nature of the relation established between sign and object. It can be characterised as a sort of 'mirroring relation' based on similarity intended widely as embedding analogy (or diagrammatic similarity) and metaphorical thinking.

As summarised in Ciula and Eide (2017: i34–i35), the practice of modelling in the sciences and as general strategy of practical reasoning has been fruitfully related with the semiotic concept of similarity and more extensively iconicity:

> Kralemann and Lattmann (2013: 3399–400) claim that models are icons, because the dominant relation with the objects they represent is one of similarity, as shown in Fig. 2. In Peircean theory, such iconic relation of similarity is what makes icons signify; icons act as signs based on how the relation of similarity is enacted: via simple qualities of their own in case of images, via analogous relations between parts and whole and among parts in the case of diagrams, and via parallelism of qualities with something else in the case of metaphors (Olteanu 2015, pp. 77, 193). […] In Kralemann and Lattmann's theory as well as in Peirce's original theory, models do not act as signs in virtue of themselves. What establishes the model as a sign is the interpretative act of a subject, whether as creator or reader. The practical act of modelling connects the model to its interpretation, that is, to its specific semantic content in a given social and institutional context (Kralemann and Lattmann 2013, pp. 3402–3). The modeller's judgement depends on his or her presuppositions connected to 'theory, language or cultural practice' (Kralemann and Lattmann 2013, p. 3417). Models are contingent. The relationship of iconicity between the model and the object being modelled is partly externally determined (it relies on the similarity between the model and the object) and partly internally determined (it depends on theory, languages, conventions, scholarly tradition, etc.). Based on this duality, they stress, on the one hand, the subjectively determined dependency of models on prior knowledge and theory and, on the other, their independence from these in light of the specific conditions of the objects being modelled (Fig. 3.1).

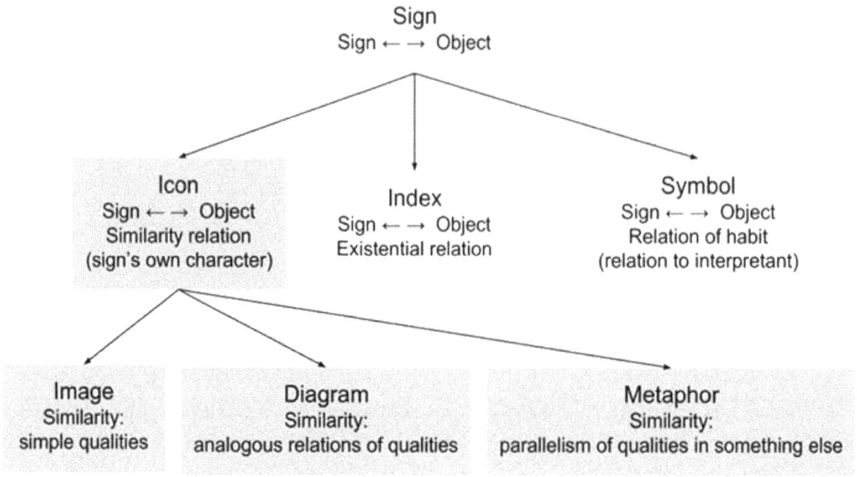

Fig. 3.1 This is a reproduction of Ciula and Eide (2017, Fig. 2), based on Kralemann and Lattmann (2013, Fig. 2). Highlighted in grey in Fig. 1 is the Peircean sign type of icon associated with models by Kralemann and Lattmann. Based on how the respective similarity relations signify, pure icons or hypoicons are classified further by Peirce into images, diagrams and metaphors. (For a recent detailed and comprehensive overview of Peirce's categories and taxonomy of signs see Olteanu (2015: 61–79).)

If, as explained by Elleström, we accept that Peircean iconicity (in the form of images, diagrams and metaphors) expresses a *continuum* of interpretative engagements from the immediacy of resemblance to highly creative cognitive leaps (Haley 1988: 34), the grades of iconicity can function as additional operative semiotic components to put to use:

> The representamen of an image is perceptually close to its object, which means that the object may be sensuously perceived in much the same way as the representamen (this is a conception that is close to Peirce's own few remarks on the image). The representamen of a metaphor is at a greater distance from its object, which means that the interpretation of a metaphor includes one or several cognitive leaps that make the similarity between representamen and object apparent. (Elleström 2013: 104)

In addition, as discussed in Sect. 3.2.3, and in line with Ellström's position (Peirce himself did not write much about metaphors as opposed to his lengthy discussions on images and diagrams), it should be stressed that metaphors do not operate only on the basis of similarity; although in cognitive and semiotic terms, icons and metaphors share a common property of signification, namely, that they are 'motivated' by similarity.

Metaphors imply systematic projection of language, imagery and inferential structure between domains/object and their/its *representamen*. According to the 'conceptual metaphor theory' (Lakoff and Johnson 1980), a 'mapping' can be identified between the conceptual domain from which metaphorical expressions are drawn (source) and the conceptual domain to be grasped via the use of the metaphor (target). However, the aim of our definition and use of metaphors in modelling in DH is to go further and surpass the dichotomy between target and source. A metaphor in fact makes emerge some constitutive elements of the object under study, namely its 'representing/translating/interpreting/vehiculating' elements that are not pre-existing; this is what constitutes the heuristic function of metaphors. The mapping process involved is closer to analogical reasoning than to similarity in the strict sense; it is mainly dependent on the entrenched relationships between object and *representamen* and not on the unidirectional correspondence of elements and features.[8] In this frame it is important not to treat metaphor and iconicity separately and, at the same time, not to dismiss their intrinsic differences. Metaphor and iconicity are closely linked, and iconicity cannot be relegated to a subordinate role confined to a mimetic relationship.

3.3.2 DH Examples

When contextualised within modelling practices in DH, iconicity and its associated wide-spectrum of graded reasoning engagements seems particularly productive:

> The notion of iconicity is however not only about how models (as signs) appear with respect to similarity to their objects. It also encompasses the possibility of manipulating models and reasoning with them. (Ciula and Eide 2017: i35)

> In general one could say that every DH model is a diagram in that it is a formalism of logical and mostly mathematical nature […]. However, we believe we can in fact identify different grades of iconicity corresponding to the three model types mentioned above, namely image, diagram, and metaphor. The classic example that comes to mind to represent an image-like model is a 3D graphic model such as, for instance, the model of an historical monument. The digital model acts as a surrogate of or a substitute for the reconstruction of the real object. A diagrammatic version of the same model could be the mathematical equations used to create the graphical 3D model. (Ciula and Eide 2017: i39)

The specific examples on which a more detailed analysis was carried out in Ciula and Eide (2017) were drawn. Respectively, from digital palaeography research for the image-like model, from research on generating digital maps from historical texts for the relational or structural (i.e. diagrammatic) model, and from a fictitious example of modelling networks of characters featuring in art historical objects for the metaphor-like model. While in this paper we focus on a wider example connected to digital prosopography, it is important to sketch out briefly how reasoning approaches could be analysed in the three cases:

---

[8]See how metaphorical mapping is modelled in the Conceptual Integration Network (CIN), developed by Fauconnier and Turner (1998); a brief but clear discussion of the two perspectives, that of Lakoff and Johnson and that of Fauconnier and Turner is available here: http://markturner.org/blendaphor.html (accessed 29 June 2017).

- Morphing reasoning. In the case of the image-like model, the model was an average morphing letter generated out of a semi-automatic process of selection of specific morphological features of segmented letter forms of medieval manuscript image samples. Inferences on the manuscript handwriting with respect to dating and localisation were based on the analysis of the morphing letter-models in virtue of an 'immediate resemblance' between the original letters in the manuscripts and the model.
- Corresponding reasoning. In the case of the relational model, relational textual expressions (e.g. location A is north of location B) were modelled into geometrical relations on their turn rendered into geographical maps. Inferences on space as expressed in the text were drawn in virtue of the corresponding spatial structure in the map.
- Metaphorical reasoning. In the case of the metaphor-like model, person names and their relationships as referred to by an art historical object (a painting in the example used in Ciula and Eide 2017) were discussed as if they were modelled respectively into entities (nodes) and into properties connecting them (links) in a network graph. As it happens in real DH models of social networks, the metaphor-like model of the net was used both to model the morphology of the painting (e.g. B's hand is laying on A's hand), the inferences on social relations (e.g. A is the wife of B) as well as the assertions of co-reference about historical people (e.g. A in painting X is the same person as C in the historical document Y) showcasing the power of metaphorical models to extend creatively across domains but potentially also the risk of applying them inaccurately to express propositions spanning different contexts and semantics.

Our conclusions stated that:

> Common for all three types of models is the inferential power operating at the interplay between their 'intrinsic structure' and their 'extrinsic mapping' (Kralemann and Lattmann 2013, p. 3409). Indeed, the features being selected in the modelling process are influenced by contextual elements of different kinds, including hypothesis, scholarly methods and conventions, sample selection, and the technologies being used. However, the inferential and epistemic power of the model relies both on extrinsic and intrinsic aspects of the model relation. In the former case, examples show us how—sometimes with vivid immediacy—similarity of existing verifiable qualities between object and model enable DH modellers to manipulate models to make new sense of those objects. In the latter case, examples show us again how models are conductive to new meaning and further modelling through our exercising of a certain imaginative freedom in selecting salient qualities and associating concepts. (Ciula and Eide 2017: i42)

3.3.2.1 The Case of the Factoid Model

A further example originally reflected upon in Ciula and Marras (2016) and further elaborated here comes from DH modelling applied to prosopography, an historical methodology for the study of pre-modern societies to collect systematically and analyse information about individual persons as attested in disparate historical sources. "Factoid" is the name associated to the prosopographical model used in several DH projects based at King's College London since 1995 and elsewhere.[9] The name evidently mirrors the "historian's worry" (Bradley and Short 2005: 8) that what they record as assertions in the historical sources under study are not the same as facts. A factoid is an assertion made by historians (in a DH context usually a project team) that a source 'S' at location 'L' states something ('F') about person 'P'. It reflects a context-aware approach to history used in prosopographical projects spanning long periods of history. With appropriate extensions, the one of the factoid has been a reusable concept successfully operationalized in DH projects. This was possible despite the fact that the historical narratives around which prosopographical studies are built are very nuanced (multidimensional, complex, non-linear), hence not easily translatable to the unambiguous language of databases:

> Multidimensionality, complexity, and non-linearity are just a few among the many characteristic features of narratives that could not be easily reduced to the unambiguous abstract language of databases. […] the factoid model has been used in a number of prosopographical projects that, taken as a whole, span across almost two thousand years of history. This would not have been possible unless this conceptual framework was general enough to allow this degree of reusability; however, it is also true that each single project required a number of extensions to the model. […] our factoid approach can show that formal structuring if designed correctly need not impose, as Veltman implies, a single perspective on the data it models, but is capable of accommodating a range of views from the different sources. (Pasin and Bradley 2015: 89–96)

Via its operationalization, the factoid concept embeds the historical approach being used but also structures the digital resources being produced. The data structure of the factoid model is operationalized as the linking of different sorts of entities (e.g. persons, places, sources, possessions etc. as schematised here http://factoid-dighum.kcl.ac.uk/what-is-factoid-prosopography-all-about/) and was tailored to several projects cases (e.g. http://charlemagneseurope.ac.uk/blog/building-a-charter-database-1-the-factoid-model-and-its-discontents/). The factoid model can be considered a pragmatic model which structures the digital resources it originates at least at the following three levels (see Pasin and Bradley 2015):

(1) Data acquisition, by acting as a guiding metaphor to conceptualise a broader historical approach outlined above and the corresponding data entry;
(2) Data storage, by acting as a practical, flexible and sustainable schema for designing databases;
(3) Data presentation, by acting as a notion around which to build user interfaces.

However, rather than a systematic application of formal structures to a specific knowledge domain (in this case the prosopography of specific pre-modern societies), according to Pasin and Bradley (2015), the factoid model can rather be considered as a process of conceptualisation and formal structuring designed to

[9]For a detailed discussion on the digital model by its main developer, John Bradley, and access to the associated semantic web ontology, see http://factoid-dighum.kcl.ac.uk/ (accessed 29 June 2017)

accommodate a range of views on a certain society from the perspectives of different sources.

DH scholarship requires adaptable models like the one of the factoid to grasp domain-specific concepts. These models stem from the specificity of DH theories and objects of analysis, more or less organically, during the modelling process, in connection to but not as effect of the observational contexts in which they are applied (Ciula and Marras 2016).

In highly computational settings, the intrinsically pragmatic component of the DH modeller's use of language is what makes the objects generated out of the application of formal and functional languages interpretable in the first place. The act of interpretation is also an act of translation (*transferre*) implying an act of 'transportation' from a language to another; from one domain and context to another; from a culture to another. In the perspective we adopt here, the concept of translation is the 'engine' of the modelling process. It allows us to overcome the dichotomy between formal and informal, object and theory, physical and mental. The process of modelling in DH is a translation between domains, it suggests to reconsider the notions of faithfulness, adequacy, and equivalence. It contributes to opening to a non-dualistic perspective, to a de-dichotomization of the polarities between object and subject, *explanans* and *explanandum*, observer and *observatur*.

3.4 Pragmatising the Modelling Process: An Interdisciplinary Approach

Our hypothesis is to propose and develop a notion of "pragmatic modelling" based upon the conceptualisation of models as icons (in the Peircean sense and as elaborated by Kralemann and Lattmann 2013) (Sect. 3.2.3) and implying the central role of metaphorical languages (Sect. 3.3.2). A semiotic conceptualisation of modelling is important to contextualise modelling not only within a framework of communication but rather of reasoning in the sense of practical thinking; as stated in Ciula and Eide (2017) this allows us to offer a new interdisciplinary perspective on how DH modelling works both in the sense of construction (how models as signs are made) as well as with respect to their epistemic value (how something new is discovered in the process of using models as signs). Within this perspective, any model is primarily pragmatic because it privileges the specificity of the objects and the contexts of use (of the humanities), but also because its pivot lies in the manipulability of the models themselves, in terms of their negotiability and flexibility (Verschueren 2012).

The focus of this approach is on the use of language in the context in which object and subject/observer operate and on the process of interpretation as a process of 'translation'. In this perspective theory and object are complementary and not associated mechanically; models therefore take a *middle position* (Ciula and Marras 2016). They are located at the convergence of multiple axes: they are contingent on social practice and on language in use (pragmatics); they imply an inductive method (bottom up) complementary to a deductive one (top down); they preserve their representational characteristics *vis à vis* the object (intentionality and extensionality); they embed the creative and imaginative power of metaphorical languages since the conceptually organised reality resulting from the modelling process is not based in a purely descriptive language and in a mere projection or interpretation of properties. Models assume a sort of "integrative function"[10] in that they offer a representation of the essential features of the object, and in that they describe the object, but also incorporate or generate possible explanations, and predictions. In a DH context the model has its syntax, grammar, and semantics but within a processual consideration of use of language that is not purely functional, but, as explained above, metaphorical, i.e. capable of adaptability (with fuzzy borders and featuring a large semantic spectrum), negotiability (the process of idealisation and simulation characteristic of mathematical models lives within a wider experiential context where rule-based and story-based models tend to co-exist). Any model can only be but a partial mapping between the model and the object being modelled, otherwise modelling would merely result in a duplication of the objects or phenomena under study, rather than in a model as intended here.

Beyond the unpacking of the similarity relation between object and model, it is important to highlight here how the components (pragmatics) of the act of modelling (contextualization, languages, actors…) in the DH context operate dynamically and consequently also how the structural polarities of object versus model can be overcome. This sense of dynamicity can be traced back to the discussion on semiotic categories borrowed from Charles S. Peirce: the world is a 'dynamic object' that is experienced in science (and in life) through the 'immediate objects' that manifest themselves as signs. Therefore, adopting an interdisciplinary perspective, we can talk of models as 'mediated objects', mediated by the conditions and constraints from which perceptions, as well as the language that expresses them, derive.

3.5 Conclusions

In this paper we tried to answer the following basic questions: Do models work as icons? Can we identify any productive metaphors used as part of modelling in DH research? What are the avenues (terminology, methods…) to overcome the apparently implicit dualism between observer and *observatur*, sign and object, subject and object?[11]

---

[10]This is also the suggestion made by the physicist Giorgio Careri in his contribution to an interesting and multidisciplinary discussion on the role of models in thought and knowledge (A.A. V.V. 1999: 185).

[11]However other questions remain open: are there other productive semiotic concepts to include in this discussion? How to unpack the relation between modelling versus inscriptions, signs versus representations?

To address these questions our suggestion is to follow two complementary paths: one related to modelling as a way of reasoning in practice in life, in science and in particular in DH (Sect. 3.2); the other strictly semiotic, considering models as icons and hence as a way of reasoning expressed in formal terms across practices and contexts (Sect. 3.3). To this aim we related dialectically DH and semiotics to discuss three central concepts, namely iconicity, modelling, and translation. Among other things, this allowed us to contextualise DH practices within a semiotic conceptualisation, which pictures modelling primarily as a strategy to make sense (signification) via practical thinking (creating and manipulating models).

By reflecting on some DH examples of modelling we claimed that a semiotic understanding of modelling could ultimately allow us to surpass the duality object versus model (as well as sign vs. context). Semiotic concepts applied and used in a DH context acquire a specific 'meaning' and a semiotic approach gives high prominence to a dynamic view on models and modelling which reinstates in renewed terms the understanding of modelling as open process. Furthermore, a semiotic approach helps in combining a "techno-scientific" perspective with a "humanistic" one.

In our paper we brought to the fore the concept of metaphorical language as a central element in modelling processes capable of accounting for the openness and the complexity of the objects being modelled; at the same time, we stressed the pragmatic nature of the act of modelling, because based on use-oriented models and because anchored to specific contexts. So formulated, the argument led to the conceptualisation of "pragmatic modelling" (Sect. 3.4): modelling is a dynamic and creative process of thinking and reasoning where meaning is negotiated through the creation and manipulation of external representations combined with an imaginative use of formal (rule-based) and informal (story-based) languages adequate to the interdisciplinary perspective within which DH operate.

References

A.A. V.V. 1999. *Il ruolo del modello nella scienza e nel sapere.* Contributi del centro linceo interdisciplinare 'Beniamino Segre' n. 100. Roma: Accademia Nazionale dei Lincei.

Black, Max. 1962. *Models and metaphors: Studies in language and philosophy.* Ithaca: Cornell University Press.

Bradley, John, and Harold Short. 2005. Texts into databases: The evolving field of new-style prosopography. *Literary and Linguistic Computing* 20 (Suppl): 3–24.

Buzzetti, Dino. 2002. Digital representation and the text model. *New Literary History* 33: 61–88.

Cecire, Natalia. 2011. When digital humanities was in vogue. *Journal of Digital Humanities* 1 (1). http://journalofdigitalhumanities.org/1-1/when-digital-humanities-was-in-vogue-by-natalia-cecire/. Accessed 13 July 2017.

Ciula, Arianna. 2017a. Modelling Textuality: A Material Culture Framework. In *Advances in 44: Papers presented at the DiXiT conferences in The Hague, Cologne, and Antwerp*, eds. Peter Boot, Anna Cappellotto, Wout Dillen, Franz Fischer, Aodhán Kelly, Andreas Mertgens, Anna-Maria Sichani, Elena Spadini and Dirk van Hulle. Leiden: Sidestone Press.

Ciula, Arianna. 2017b. Digital palaeography: What is digital about it? *Digital Scholarship in the Humanities* 32 (suppl_2.1): ii89–ii105.

Ciula, Arianna, and Cristina Marras. 2016. Circling around texts and language: towards "pragmatic modelling" in digital humanities. *Digital Humanities Quarterly* 10 (3). http://www.digitalhumanities.org/dhq/vol/10/3/000258/000258.html. Accessed 13 July 2017.

Ciula, Arianna, and Øyvind Eide. 2014. Reflections on cultural heritage and digital humanities: Modelling in practice and theory. In *Proceedings of the First International Conference on Digital Access to Textual Cultural Heritage* (DATeCH'14). 35–41. New York: ACM. http://dx.doi.org/10.1145/2595188.2595207.

Ciula, Arianna, and Øyvind Eide. 2017. Modelling in digital humanities: Signs in context. *Digital Scholarship in the Humanities* 32 (Suppl. 1): i33–i46. https://doi.org/10.1093/llc/fqw045.

Daston, Lorraine (ed.). 2000. *Biographies of scientific objects*. Chicago: The University Chicago Press.

Elleström, Lars. 2013. Spatiotemporal aspects of iconicity. In *Iconic investigations*, ed. Lars Elleström, Olga Fischer, and Christina Ljungberg, 95–117. Amsterdam: John Benjamins Publishing Company.

Fauconnier, Gilles, and Mark Turner. 1998. Conceptual integration networks. *Cognitive Science* 22 (2): 133–187.

Flanders, Julia. 2009. The productive unease of 21st-century digital scholarship. *Digital Humanities Quarterly* 3. http://www.digitalhumanities.org/dhq/vol/3/3/000055/000055.html. Accessed 13 July 2017.

Flanders, Julia. 2012. *Modeling scholarship*. Paper presented at the workshop Knowledge Organization and Data Modeling in the Humanities: An ongoing conversation, Brown University, RI, March, 2012. https://datasymposium.wordpress.com/flanders/. Accessed 13 July 2017.

Flanders, Julia and Jannidis, Fotis. 2015. *Knowledge organization and data modeling in the humanities*. [White Paper]. https://opus.bibliothek.uni-wuerzburg.de/frontdoor/index/index/docId/11127. Accessed 29 June 2017.

Geertz, Clifford. 1973. *The interpretation of culture*. New York: Basic Books. Reprint: Geertz, Clifford. 1993. *The interpretation of culture*. London: Fontana Press.

Graves-Brown, and Paul M. 2000. Introduction. In *Matter, materiality and modern culture*, ed. Paul M. Graves-Brown. 1–9. London: Routledge.

Haken, Hermann, Anders Karlqvist, and Uno Svedin (eds.). 1993. *The machine as metaphor and tool*. Berlin, Heidelberg: Springer.

Haley Michael C. 1988. *The semeiosis of poetic metaphor*. Bloomington, Ind.: Indiana University Press.

Hesse, Mary. 1966. *Models and analogies in science*. Notre Dame: University of Notre Dame Press.

Kralemann, Björn, and Claas Lattmann. 2013. Models as icons: Modeling models in the semiotic framework of Peirce's theory of signs. *Synthese* 190 (16): 3397–3420.

Lakoff, George, and Mark Johnson. 1980. *Metaphors we live by*. Chicago: University of Chicago Press.

Mahr, Bernd. 2009. Information science and the logic of models. *Software & Systems Modeling* 8: 365–383.

Marras, Cristina. 2006. *Metafore scientifiche, entry for the Enciclopedia Filosofica Italiana*, ed. Virginio Melchiorre. 7364–7366. Milano: Bompiani.

Marras, Cristina. 2013. Structuring multidisciplinary knowledge: Aquatic and terrestrial metaphors. *Knowledge Organization* 40 (6): 392–399.

Marras, Cristina. 2014. Exploring digital environments for research in philosophy. Metaphorical models and sustainability. In *AIUCD '14, Proceedings of the Third AIUCD Annual Conference on Humanities and Their Methods in the Digital Ecosystem*, Bologna, September 18–19 2014, eds. Francesca Tomasi, Roberto Rosselli del Turco, Anna Maria Tammaro. ACM, NY. https://doi.org/10.1145/2802612.2802639.

Marras, Cristina. 2017. *Les métaphores dans la philosophie de Leibniz*. Limoges: Lambert–Lucas.

McCarty, Willard. 2005. *Humanities Computing*. Basingstoke [England]; New York: Palgrave Macmillan.

McCarty, Willard. 2006. Tree, turf, centre, archipelago–or wild acre? Metaphors and stories for humanities computing. *Literary and Linguistic Computing* 21 (1): 1–13.

McCarty, Willard. 2009. Being reborn: The humanities, computing and styles of scientific reasoning. *New Technology in Medieval and Renaissance Studies* 1: 1–23.

McCarty, Willard. 2014. Getting there from here. Remembering the future of digital humanities Roberto Busa award lecture 2013. *Literary and Linguistic Computing* 29 (3): 283–306.

McGann, Jerome. 2014. *New republic of letters: Memory and scholarship in the age of digital reproduction*. Cambridge, MA, USA: Harvard University Press.

Moretti, Franco. 2013. Operationalizing. *New Left Review* 84: 103–119.

Morgan, Mary S, and Tarja Knuuttila. 2012. Models and modelling in economics. In *Philosophy of Economics*, ed. Uskali Mäki. Handbook of the Philosophy of Science, eds. Dov M. Gabbay, Paul Thagard, and John Woods. 49–87. Elsevier Science.

Nersessian, Nancy J. 2008. *Creating scientific concepts*. Cambridge, MA: MIT Press.

Olteanu, Alin. 2015. *Philosophy of education in the semiotics of Charles Peirce. A Cosmology of Learning and Loving*. Oxford, Bern, Berlin, Bruxelles, Frankfurt am Main, New York, Wien, Peter Lang.

Ortony, Andrew (ed.). 1993. *Metaphor and thought*. Cambridge: Cambridge University Press.

Palmieri, Fabio (ed.). 2012. *Consciousness in interaction. The role of the natural and social context in shaping consciousness*. Amsterdam: John Benjamins Publishing Company.

Pasin, Michele, and John Bradley. 2015. Factoid-based prosopography and computer ontologies: Towards an integrated approach. *Digital Scholarship in the Humanities* 30 (1): 86–97. published online June 29, 2013 https://doi.org/10.1093/llc/fqt037.

Peirce, Charles, Sanders. 1933. *Collected papers of Charles Sanders Peirce* [CP], eds. C. Hartshorne, P. Weiss. Cambridge, MA: Harvard University Press.

Small, Helen. 2013. *The value of the humanities*. Oxford: Oxford University Press.

Verschueren, Jef. 2012. The pragmatic perspective. In *Handbook of pragmatics*, ed. Jef Verschueren, Jan-Ola Östman, and Jan Blommaert. Amsterdam: John Benjamins Publishing Company.

Chapter 4
Rationality and Reasonableness in Textual Interpretation

Massimo Leone

> *Wise men have interpreted dreams, and the gods have laughed.*
> H. P. Lovecraft, *Hypnos* (1922).

Abstract What is the difference between interpreting a literary text during a university lecture and interpreting reality outside of the academe? And what is the difference between interpreting in natural sciences and interpreting in the humanities? Despite evident and known divergences, humanities too can rank their interpretations and aspire to guide the interpretations of society. Three alternative methods can be used so as to test interpretive hypotheses, depending on whether the author's, the reader's, or the text's meaningful intentionality is primarily investigated. The third method is superior to the first two since it leads to the creation of a common meta-discursive space for inter-subjective exchange about meaning. Although adopting an appropriate methodology is essential in textual analysis, that which is even more important is supporting the creation of a community of interpreters that, sharing the same method, engage in the constructive comparison and ranking of interpretive moves.

4.1 Introduction: The Reluctant Interpreter

If everyone were spontaneously inclined to interpret reality, be it nature or culture, in the same way, interpretation would not be needed. It would be pleonastic. In the relation between two different natural languages, an interpreter is necessary for she or he masters the language of departure and that of arrival better than any of the speakers of the two languages. That is why she or he is not only needed but also somehow admired for this capacity, which inevitably leads to power (Fawcett et al. (2010), part III and 147, 150–152). Analogously, the interpret of a text is someone who, thanks to a superior methodological training, a wider encyclopedia, and a better acquaintance with the text, can provide insights on its meaning that are to be judged as deeper and more truthful than those that the interpreter's audience might

M. Leone (✉)
University of Turin, Turin, Italy
e-mail: massimo.leone@unito.it

A. Olteanu et al. (eds.), *Meanings & Co.*, Numanities - Arts and Humanities in Progress 6, https://doi.org/10.1007/978-3-319-91986-7_4

bring about. In other words, interpretation begets intermediation, but intermediation is inseparable from an idea of hierarchy. The interpreter, for the fact itself of being an interpreter, will be situated in a higher sociocultural and pragmatic level than the community to which she or he addresses her- or himself. That is evident in the whole history of interpretation in the western civilization (Barnstone 1993). In the Jewish Bible, for instance, all those who are able to read the signs of transcendence better than their fellow human beings acquire a special status, often becoming the leaders of the community. The same happens to prophets and haruspices in the Jewish, Greek, and Roman culture (Struck 2016: 48, 221, 223). In a similar way, when the professor of semiotics addresses the audience of her or his students, and undertakes the task of describing, analyzing, and interpreting a text—a painting, for instance—immediately a topology of symbolical power materializes in the classroom, having often as its concrete counterpart also a spatial topology (the professor at the center of the amphitheater, the students all around) (Oblinger 2006: Chaps. 1 and 6).

Lectures of this kind, however, constantly reach the bittersweet moment when the professor finally concludes the analysis and shows, with masterful assurance, that the meaning of a novel, a painting, or a song can be encompassed in such and such sentences of the meta-language. Through the rigorous application of a method, and the careful consideration of arguments and counterarguments, the analyst reaches the conclusion that the meaning of a text can be only that expressed through the analysis. Like in the final stage of the solution of a puzzle, all the pieces seem to come together and to find the most suitable place in the organic gestalt of the analysis (Danesi 2002: Chap. 7). When the professor then invites the audience to ask questions, a couple of students might propose alternatives for the interpretation of such or such detail, or even point out that some aspects of the text were not taken into sufficient consideration. The masterful professor, though, will not be confounded by such objections but will welcome those additions that fit well within the general design of the interpretation and will, simultaneously, calmly but vigorously argue against alternatives that would disrupt the whole hermeneutic picture. At the end of the lecture, and even more at the end of the session of questions and answers, the interpretation will stand like a Doric temple in front of the minds of the audience, like a conceptual architecture wherein nothing, not even the smallest element of decoration, is now missing.

On the one hand, this moment is sweet: the professor will have demonstrated her or his superior ability to decode culture and its texts, to extract the most coherent patterns of meaning from them, and to translate such meaning into a smooth meta-language, enhancing the students' comprehension of what the Divine Comedy, or the Sistine Chapel, or the Fifth Symphony of Beethoven actually mean. Yet, to the professor who will not fall completely and blindly prey of the inebriation that accompanies teaching and, more generally, any display of a supposedly superior ability to interpret, this apex of the analysis will have a bitter taste too. The professor will mentally look at the tetragonal façade of the analysis, behold its cold splendor, but then she or he will inevitably be seized, in a moment of sincerity, by a tragic doubt: what if this is all wrong? What if I am just trying to convince myself, and my students together with myself, that meaning is something that can be captured once and for all, transcribed into a formula, transmitted across the ages and the generations, agreed upon?

The 'reluctant interpreter' will look at the poem, at the painting, at the song, and wonder whether that which lies in front of her or his eyes is not a living animal anymore but a prey from whose corpse the last sparkles of life are nervously fleeting away. At the same time, she or he will not be able to suppress the intuition that, under that beautiful but cold architecture, the swarming life of meaning is still going on, undetected by anyone except those who have not been entirely convinced by the professor, perhaps including the professor her- or himself. At the end of the analysis, indeed, only the arrogantly distracted, pompously boastful interpreter will be able to tacit that inner voice that stealthily suggests that the beautiful architecture is not made out of marble but out of clay. Are the students really convinced, or are they just bored and eager to abandon the classroom? More disquietingly, do they accept the professor's interpretation because it is objectively the best, or because they rather yield to the professor's authority, to the social habit of respecting, or even fearing, her or his hierarchically superior level of judgment? Even more puzzlingly, the professor will ask her- or himself whether she or he too is actually not convinced either, and is simply faking to be persuaded of the soundness of the interpretive solution, for that is what the role of the professor implies; that is what the role of the interpreter implies. She or he perhaps does not fully understand what is being said or what has been written, but her or his role imposes the pantomime of understanding.

4.2 The Big Gap

What causes this excruciating doubt, however, is not the fragility of the method, or the suspect of the students' insincerity, or the idea that out there, elsewhere, some other professor might interpret the text in a better way. What causes the doubt, instead, is life. Life itself. The reluctant interpreter, indeed, will not be able to discard the sneaking idea that there is a close and poisonous relation between the practice of interpretation taking place in vitro, that is, in the artificial atmosphere created in the laboratory of a classroom, and the practices of interpretation that, instead, occur spontaneously all around the academic world, and that in most cases lead to tragedy and disaster. Is the approach through which a poem is analyzed perhaps so different from the approach that, day after day, leads a human being to completely misunderstand a friend, or to blindly fail to notice a partner's infidelity, or to let the lies of a politician go totally undetected, or to be spectacularly mistaken about the future directions of society or economy? Every day, all around the world, people suffer or even die because they have wrongly interpreted the environment in which they were immersed (Reason 2013: 10). That is why only the foolish human being will not have, at some stage of her or his life, the tremendous thought: I don't understand anything; I have never understood anything; I had the impression that

I was able to decode life and reality but I was not; it was just an illusion, whose tragic price will be paid all of a sudden, and all at once.

The contrast between what happens to interpretations in a classroom and what happens to them in life 'out there' is so striking that it must, at a certain stage, prompt a reflection on the status of the former in relation to the latter. The reassuring pedagogical vulgate of academic teaching is that education, and higher education in particular, exactly is that physical and conceptual place in which young minds, guided by more experienced minds, can learn how to produce those interpretations that will improve the quality of their life and that of other people. Young biologists, young doctors, young psychologists, young semioticians are familiarized with heuristic strategies and tactics that have proven effective in the past, and that are likely to prove effective in the future as well. Members of a cognitive elite, students will guide the rest of the population in its daily interpretations, influencing the way in which people build their houses, educate their children, cure their diseases, choose their novels, talk about their private life, and interact with other human beings. In other words, that position of hierarchically superior interpretive status that the professor currently holds in front of them will be the same that they themselves will hold in their social and professional environments, for the simple fact that they were exposed to and absorbed a series of methodological habits, whose mastery will be certified by their diplomas.

Why, then, is the passage from the classroom to the street, or to the square, or to the world, experienced as increasingly painful by the majority of students? Why do more and more of them believe that what they have learned from university lectures does not prepare them at all to deal with what they will face as adult architects, or physicians, or lawyers? Is it just a matter of updating the information that students receive during their training? Or filling the gap between university and 'reality' through multiplying those experiences that students can gain across these two worlds, such as stages or internships? Will a generation of younger professors help in refreshing and sharpening the interpretive skills of students? Unfortunately, the semiotician that looks with satisfaction at the complex diagram through which she or he is explaining to students the meaning of a poem, or a photograph, or a theatre performance, will have to admit that the problem is not as simple as that. She or he might, of course, update the methodology, read the latest books in the field, attend symposia that deal with the most current developments of the discipline, etc. The feeling of dismay at the thought of the enormous distance between the epistemological atmosphere of the classroom and that of the world is destined to persist, or even to be embittered by the experience of how useless any effort of methodological upgrade turns out.

4.3 Methods of Corroboration

The feeling of this distance, which is at the same time a cognitive, an emotional, and a pragmatic one, has to do with some central issues not only in semiotics, but also in any other discipline involving a systematic reflection on interpretation. Such issues affect the jurist as well as the medical scientist, the art curator as well as the journalist (Maitland 2017: "Introduction"). In a nutshell, such distance can be measured in relation to the following question: what are the acceptable procedures of interpretation that are usually adopted during a university lecture, and what are, on the contrary, those that prevail outside of the academic laboratory? In order to answer such question, one should posit an epistemological, methodological, and analytical distinction between rationality and reasonableness. As it has been pointed out by most modern philosophy of science, a great divide separates the procedures through which natural scientists reach their conclusions through interpreting the data deriving from the observation and analysis of reality and the procedures through which 'human scientists' do the same with both sociocultural reality and the texts that it produces (Popper 1935, Chapter 4: "Grade der Prüfbarkeit"). In the first case, the replicability of the experiment allows scientists to test hypotheses in order to either corroborate them or falsify them, so that they might be replaced by better, more encompassing speculations about the nature of nature.

When a semiotician describes, analyses, and interprets a text, her or his hypotheses cannot be tested in the same way. In a recent article of mine, for instance, I have claimed that one of the most famous and discussed upon sequences in the history of cinema, the "sequence of the vase" in *Late Spring* by Ozu (1949), should be interpreted by taking as a point of departure not only the famous vase that all previous interpreters had exclusively focused on, but also the strange object that appears, unnoticed by all, in a shadowy corner of the same sequence (Leone 2016) (Fig. 4.1).

I have proposed this interpretation to my Japanese colleagues, to my Turin colleagues that are experts of Japanese cinema, and, last but not least, I have discussed about my reading at length with the students of my master course of semiotics. I have also published this interpretation online and in journals, both in Italian and in Japanese. All the concerned audiences seemed to be convinced by my interpretive move, and no major objections were leveled at it. But how can this semiotic hypothesis be corroborated, and how can it be falsified? According to the theory of semantic intentionality formulated by Umberto Eco, there are three different ways to test the hypothesis above (Eco 1990, Chapter 13: "Semantics, Pragmatics, and Text Semiotics").

Fig. 4.1 Ozu's *Late Spring* (1949)

4.3.1 Asking the Author

The first is to contrast it with the semantic intentionality of the author, that is, what the author meant to signify and communicate through the sequence. This modality of verification, however, is not the most effective, for two reasons. A contingent reason is that, since Ozu is no longer alive, we cannot ask him what he meant by the sequence of the vase. He might have left some meta-texts, such as notes, interviews, etc., explaining the meaning of the sequence. One could not rely on these meta-texts, however, in the same way as one would rely on Ozu's own words. Determining in what way these meta-texts explain the meaning of the sequence, indeed, would require further interpretations, some of them being as, if not even more, speculative than the one seeking to interpret the sequence in the first instance. The second reason for the unviability of this procedure is that Ozu himself cannot be regarded as the highest possible authority on the meaning of his movies. He might have had an idea of what the sequence of the vase might signify, or rather of what he wanted it to signify, and he might have tried to arrange all the elements of the sequence so as to signify it. Given the nature of the cinematic language, however, which does nothing but magnifying the nature of language tout court (Aumont 1990, Chapter 6: "Les pouvoirs de l'image"), the sequence itself signifies despite and beyond Ozu's intentions. There is no signification without the potentiality of alternative interpretations. Even those texts that are constructed so as to reduce such alternatives, like the instructions of a drug's leaflet, for instance, can generate diverging interpretations, exactly because of the intrinsically undetermined nature of verbal language (Eco 1976, Chapter 1: "Signification and Communication"). Such indeterminacy is even higher when images, instead of words, are used in order to assemble the signification of a message, especially when such images are not meant to refer to reality but to tell a story and, even more ambiguously, to convey a moral message about abstract human values. To this difficulty one should add that even if Ozu could perfectly explain to us the meaning of the sequence, the explanation would be a fragment of verbal meta-discourse and, therefore, adopt a different expressive substance, syntactic structure, and semantic behavior than the sequence itself (Calabrese 2000). Ultimately, that which Ozu might say on a sequence of one of his movies cannot be considered as its definitive interpretation because that which Ozu really wished to signify with that sequence, he signified it precisely through it, whereas, when requested to explain the same signification by translating it into a verbal meta-language, he would face the same difficulties that most critics do, and without even the benefit of detachment.

4.3.2 Asking the Reader

A second viable option to test my interpretive hypothesis might be to compare and contrast it with those of other spectators. That in a way happens every time that an interpretive hypothesis is communicated: in my case, the reading of the sequence was put to the test of its reception by Japanese colleagues, by Turin Asian cinema specialists, even by a historian of Japanese pottery, and, of course, by students. In all of these circumstances, as it was pointed out earlier, no objection was made. Is that sufficient as a means to corroborate the reading of the sequence as the best possible interpretation? The problem here arises to designate the community of interpreters that would exert the highest authority in matters of judgment over the extent to which an interpretation of the famous sequence of the vase in Ozu's *Late Spring* might or might not be correct. How and by whom is this community going to be shaped? Perhaps by Japanese spectators, since they are supposed to be those whose interpretive encyclopedia is geographically and culturally the closest to that which Ozu referred to in creating the sequence itself? But, then, how should one select and shape such cohort of Japanese spectators? What age should they be, from what sociocultural background, of what gender? The major problem in adopting a corroboration strategy based on the notion of 'intentio lectoris' (the meaning that the reader of a text, or the spectator of a movie, attributes to such cultural artifact) is its representativeness. In other words, against what kind of range of subjective interpretations should mine be tested? The answer is not clear and might even lead to solutions implying intolerable ethnocentrism. Should only Italians interpret Dante, and only the French Molière? Is the native command of a natural language enough to determine the circle of those who are supposed and entitled to correctly interpret a text? And is this applicable also in the case of predominantly non-verbal texts, such as the famous sequence of the vase in Ozu's movie? The major difficulty of this second method, however, resides in the fact that one should clearly establish beforehand the criteria according to which a subjective interpretation is going to be judged as more apt to falsify or corroborate a hypothesis than another.

Statistic methodologies would not be a solution either. Should one conclude that, let's say, since in a survey more than 75% of spectators have declared that they interpret the famous vase as symbol of the solitude of the male protagonist of the movie after the demise of his wife, this should be considered as the correct interpretation? And would this statistic argument imply that, should a subsequent survey show different statistics, the correct interpretation should change accordingly? Such way of testing, falsifying, and corroborating interpretive hypotheses would be unacceptable for most scholars in the humanities, although some of them are exploring these procedures in the field of quantitative aesthetics (Kreuz and MacNealy 1996). What is particularly problematic in this statistic methodology is that it would export a method of political determination (the majority wins) in a field where the opinion of the majority is not necessarily the lesser evil. On the contrary, one could claim that one of the objectives of the critical analysis of artworks is not to comply with the average opinions expressed by a community of receivers but rather to alter this average through showing that more complex and all-encompassing interpretive hypotheses are possible, and potentially lead to farther exploit the moral content of an artwork. In the humanities, art historians and literary critics have not been usually seen as those who certify the interpretations of the average receiver, but rather as those who, through engaging with the artwork at length, discover in it paths of meaning that the most were not able to detect.

4.3.3 Asking the Text

The third major strategy for testing an interpretive hypothesis does not consist in asking directly or indirectly the author (which, moreover, is impossible in some cases, think about the supposedly transcendent authorship of most 'sacred' texts) or in seeking to determine a majority of readers but in comparing and contrasting different interpretations in relation to the artwork itself. In our example, this heuristics consists in comparing my interpretation of the sequence of the vase with other interpretations that different spectators might have produced about the same sequence. Although this strategy seems similar to the second one, it is, indeed, a completely alternative one. In the subjective method, different interpretations are compared and contrasted, but there is no rational standard for doing it, save for that of ethnocentric or similar conventions (someone is Italian, so her or his interpretation of Dante must be better; someone is a Japanese man, so his interpretation must be better than that of a Japanese woman), or that of statistic measurements. In the third way, instead, different interpretations are compared and contrasted not subjectively but inter-subjectively, not in relation to a supposed primacy of a certain sociocultural category of readers or to that of the majority, but in relation to the text itself. That is why this method leads to a rational comparison of interpretations:

the way in which they differ can be articulated and measured in relation to the text, that is, to features of the texts that are visible to everybody. To give an example, one thing is to say: "let us compare the Japanese interpretations of Ozu with the Italian ones", which is a comparison that makes reference to extra-textual and quite slippery elements, such as the nationality of spectators; another thing is to say: "let us compare the interpretations of the sequence that only consider the vase, with those that consider both the vase and the mysterious object appearing in a shadowy corner of the scene"; or, "let us compare the interpretations that explain why the vase is shown twice in the sequence, to the interpretations that do not".

In relation to the first two methods, this third one implies a significant advantage, which is the possibility of rationally ranking interpretations. In the first method of testing, there was no particular reason to believe that Ozu's interpretation of his own sequence was superior to those of his spectators, unless one relied on a naïve idea of signification, an idea whose weaknesses have been exposed supra; similarly, there was no particular rational argument to believe that an interpretation be superior to the others because of its ethnocentric or statistic force. The third method, instead, introduces a simple but irrefutable logic of ranking: since the whole of the text signifies, those interpretations coherently taking into account more elements of this signification will be preferable to those that take into account only a part of them (Hjelmslev and Uldall 1957). That is an axiom, but it is one that is difficult to refute. It rests, indeed, on the concept itself of semiosis. In verbal language, for instance, given a sentence, or even a complex assemblage of sentences, such as a literary text, it is hard to demonstrate that an interpretation that takes into account only its incipit is better than that which, on the contrary, describes, analyses, and interprets the totality of verbal signs woven within the text. Similarly, it is hard to prove that a reading of a painting of Piero della Francesca explaining only the meaning of the characters appearing therein but not that of the background is preferable to an interpretation where both receive an analytical reading. This axiom also relies on the principle that culture is made not only of texts but also of para-texts, that is, meta-signs that are supposed to tell readers (observers, listeners, etc.) where a text starts and where it ends, from where the analytical effort must be exerted and where it can stop (Genette 1987). Postmodern artists and texts can play with such para-textual meta-signs, producing, for instance, artworks whose frontiers are not neat but blur with that of their contexts. This blurring effect, however, is possible only for, in most circumstances, culture clearly signals the limits of a text. Since culture tells us how and where to look at in order to understand where and when a text begins, and where and when it ends, it is reasonable to believe that culture also requires to exert our interpretive attention throughout the text, and not only through part of it.

4.4 Interpretive Temporality

This criterion of exhaustiveness is in a way superior to any criterion adopted either in the objective or in the subjective methods of testing, because it is an intrinsically cooperative one. Whereas the first method implicitly affirms that only the author knows the meaning of a text, and whereas the second intrinsically claims that only a certain group of readers knows it, the third method leaves open the question of who might better interpret a text: it is not the author; it is not a certain category of readers; it is, on the contrary, the one or those who will look more attentively into the text and explain more carefully what they can see and read. The first method is oriented toward the past: the author conceived of a meaning, expressed it through an artwork, and it is now upon the reader to retrieve that meaning from the text, if possible by faithfully adhering to the author's first communicative intention. The second method, on the opposite, is oriented toward the present: it is here and now that the text means, for a reader x or y, and its meaning is not to be searched in the past intentions of an author—intentions that preceded the creation of the artwork—but in the current perception of a selected group of readers, whose features or number make them the artwork's best interpreters. A farther reason for which the third method of testing is preferable to the first two is that it is, instead, open to the future: an interpretation is put forward, but if in the future someone will propose an interpretation that explains all the elements taken into account by the previous interpretation, and explicates, moreover, some farther elements, then this future interpretation will have to be judged as superior.

In the first and in the second method, a better interpretation can also emerge, but not in relation to the text; in the first case, a better interpretation will be one that, for instance, relies on novel information concerning the intentions of the author, information that was hitherto ignored or neglected; in the second case, a change in the ethnocentric polarization of the semiosphere or in statistics will make a better interpretation emerge: now that women more than men interpret Hamlet, for instance, a feminist interpretation must be preferred to a macho one (on the weaknesses of this approach, Zhang 2016). The dynamic of improvement implied by the third method is intrinsically different and superior to the first two because it does not relate to extra-textual elements but to intra-textual ones; in other words, in order for someone to come up with a better interpretation, it is not necessary to look for it outside of the text, because such interpretation is already available to everyone inside of the text, and all that is needed is to better look into it. To return to the example given above, the sequence of the vase in Ozu's *Late Spring* has always contained two objects, not only the vase. It was always possible for someone to notice the second object and to include it into the interpretation. For some reason, however, this second object was neglected, until someone took notice of it and included it in a coherent interpretation of the sequence. This new interpretation, which takes into account more elements of the textual manifestation than the previous ones, is better than them not because it better grasps the intentionality of the author, or because it fits better with the inclinations of a group of readers, but simply because it explains more and better that which all might have seen and interpreted in the text.

Along this line, interpretations of King Lear that propose a coherent view of more textual elements of it are to be preferred to those interpretations that are sketchy, or partial, or neglect to fit some major portions of the text within the architecture of the interpretation. Similarly, interpretation of the Sistine Chapel that coherently read more visual details of it, without neglecting those that are placed at the margins of the visual scene, are also to be preferred. In a way, this strategy for ranking interpretations is not dissimilar from that which guides natural sciences too: if someone formulates a hypothesis that might explain more of that which can be seen in the image produced by a microscope, then that hypothesis must be preferred to that which neglects some of the elements of the image.

4.5 Textual Laboratories

Something central, however, still distinguishes the third method for testing the value of a textual interpretation from Popper's idea of scientific corroboration. All the interpreters of a text deal with the same materials. They can take into consideration a smaller or a greater quantity of them, but they cannot exit the perimeter of the paratext. Interpreters of Shakespeare's *King Lear*, for instance, can propose different readings of some of the passages of the text, but they cannot add new materials to it. In natural sciences, instead, the fact that an experiment can usually be replicated allows scientists to control the changing of conditions that surround the interpretation of reality. In a textual reading, an interpretation can improve a previous interpretation by taking into consideration neglected elements of the text, but not by adding new elements to it; in natural reading, since the experiment is supposed to be not the ultimate object of interpretation, but a controlled simulacrum of reality (nature), then its para-textual frontiers are not as clear: through varying the frontiers of the simulacrum and, therefore, its semiotic relation with reality (nature), better interpretations can be formulated, interpretations that are preferable not simply because they explain neglected elements of the experiment, but also because they introduce neglected elements into it.

Mutatis mutandis, one might affirm that a textual reading cannot be compared with a scientific reading, but rather with the designing of an experiment leading to a scientific reading, the only but crucial difference being that the humanist's experiment cannot vary, because the laboratory coincides with the object that is to be interpreted. The frustration of the humanist interpretation in relation to the scientific one derives from the fact that the textual materials that eighteenth-century scholars had at their disposal so as to interpret Shakespeare's King Lear are actually the same as those that contemporary interpreters have. Shakespeare's text is both the object of the reading and its laboratory, and the same syllables, the same words, the same sentences, and the same rhymes constitute the perimeter of that which the interpreter can read and to which she or he can attribute meaning.

If the para-textual perimeter of a text does not vary, though, in the sense that it offers always the same elements to any interpreter who might come across the significant surface of the text itself, the lenses of the microscope through which these elements are observed might indeed vary. The psychoanalyst of literature and the semiotician look into the same text, but they see different elements in it.

4.6 Interpretive Lenses

The way in which readers, spectators, listeners, etc. look at the significant surface of a text is never completely casual, or merely guided by personal taste. It is, on the contrary, shaped by a series of patterns, some of which derive from the historical and sociocultural coordinates of the reader, whereas other stem from her or his subjective training as reader and interpreter. It is known, for instance, that being an assiduous reader of the literary works of a certain author turns one into a different kind of interpreter of that author, one who is able to detect, within that author's texts, elements and nuances that less accustomed readers tend to ignore. In this case, it is not the text that varies, but the cognitive and emotional filter through which its significant surface is sifted. A novel invests the readers with a myriad of words, sentences, and stories; a movie solicits the beholder with a frantic sequence of images; etc. The novel and the movie are the same for everyone, meaning that they present all readers with the same significant surface, and yet not the entirety of it can be retained and deciphered on the spot. Exiting a movie theater, or closing a book, some more attentive and perceptive interpreters will have retained a greater amount of information, and some of them will have almost naturally conflated this information into a coherent interpretation. The value of an interpretive method, therefore, can be measured in relation to the extent to which it leads interpreters to discover, within the perimeter of a text, elements that they would not have otherwise detected. If Greimassian semiotics has a merit, for instance, exactly consists in working as a lens that, pointed at a certain text, allows one to see elements within it that would have been otherwise gone unseen and neglected (Greimas 1976).

That is possible because every interpretive method essentially consists in a translation. In the Greimassian method of textual interpretation, that which matters the most is not the exactitude of the theoretical proposal, many aspects and nuances of which are both debatable and, indeed, debated. Just to give an example, this method imagines the meaning of a text as though it was generated through a series of logical operations, whose dynamics and results can be ranked at different levels of abstractedness. Despite what Greimassian 'fundamentalists' tend to affirm, there is no intrinsic motivation for singling out a certain number of such logical levels (that of deep values, that of narrative grammar, that of discourse, and that of figuration). This largely depends on the accidents of the history of the elaboration of the semiotic method, for instance, on the fact that some aspects of the theory were developed more than others.

The major usefulness of the Greimassian method, instead, emerges from the fact itself of translating the features of a (literary, filmic, visual, etc.) text into a new analytical simulacrum, whose features coherently derive from the inter-defined meta-language that the analysis adopts. In other words, when the Greimassian method translates the meaning of a text through rendering it into a meta-language, that which matters is not as much the meta-language itself, which could be replaced by other meta-languages, but the coherence of the analytical translation. Translating the meaning of a text into the code of a meta-language, indeed, allows the analyst to see, within the perimeter determined by the text's para-textual marks, new elements and new dynamics. A lay reader will see, when reading Shakespeare's King Lear, a complex array of syllables, words, and sentences; she or he will also see a protagonist, perhaps a co-protagonist, some secondary characters, and the line of a narrative plot. When a semiotician trained in the Greimassian method looks at the same text, she or he will see the same elements, but she or he will simultaneously detect also elements that the lay reader does not perceive, exactly for they emerge from the application of the meta-language to the text itself. The Greimassian semiotician, for instance, will see not only characters, but also macro-narrative functions that complexly intertwine with them.

That is why the application of a method enhances interpretation: not in the trivial and somehow misleading sense that such method leads to the discovery of a textual truth that non-trained approaches are unable to spot, but in the sense that, through the operation of meta-linguistic translation, the method multiplies the possibilities of elaborating interpretive hypotheses concerning the text itself. As it was stated earlier, unlike in natural sciences, in human sciences, the text is both the object under investigation and the investigation laboratory; its perimeter must remain the same. The application of a new method, that is, of a new way of meta-linguistically translating the text, introduces new possibilities for the analysis. The text is always the same, but now the meta-text allows one to peruse it under a different angle.

This meta-linguistic translation is not without problems: it must be faithful to the significant surface of the text and, at the same time, render its structures in a creative way. If one looks at this matter with unbiased eyes, however, one must admit that the so-called "lay" approach to a text is not naturally unproblematic either. The fact that we recognise "sentences" in a literary text might appear as a spontaneous operation, but such spontaneity actually is the result of interiorizing a grammatical analysis whose evolution required a long and sometimes conflictive elaboration. The fact that we recognise a subject, a verb, and an object in King Lear's sentences is not less conventional than the fact that we recognise an actant subject, an actant object, or an actant observer in the narrative plot of the text. The difference between the two approaches is not as much a question of abstractedness as one of centrality, within the semiosphere, of such or such method of interpretive reading. In most contemporary societies, readers are trained to analyze the syntax of a text, not its semantics, which is usually still dealt with through impressionistic methods.

Nothing, however, intrinsically prevents a structural method of semantic analysis to become equally central in the semiosphere, to be taught and learned in schools, and to become a 'second nature' of most readers and readings.

4.7 Towards an Interpretive Liturgy

That leads one to consider the major advantages of this third method for testing the interpretive hypotheses concerning a text, which is the possibility of sharing and inhabiting a common space for inter-subjective discussion. In the first method, there is no need to debate about interpretations: those who have the closest access to the supposed mind and intentionality of the author occupy the highest hierarchical rank in the sphere of interpretation. Such method is dangerously inclined to authoritarianism, especially in those cases in which the identity of the author is not clear or is mystified. In the "religions of the book", for instance, power can be held also as a consequence of the claim that one, be it a scribe or a priest, is closer to the intentionality of the author, the problem in this case being that there is no way to prove or disprove such pretense of proximity (Leone 2013). In the second method too, discussion is not a viable option, because diverging interpretations are not meant to reconcile their differences and converge toward a negotiated version of them, adopting those tenets that better fit the reading of the text at stake, but to coagulate into statistical clusters: as long as an interpretation is shared by a majority of readers, it is to be considered the correct one, no matter what.

In the third method, instead, which is the one that targets the *intentio operis*, discussion is not only an option but is the natural outcome of how this approach works. When the meaning of a text is translated into a diagram that renders it through a specific meta-language, then competition must arise among all the beholders of such diagram, in order to determine who is able to see more elements into it and, more importantly, who is able to connect the greatest quantity of such elements into a coherent interpretation. There is no more exciting moment, in academic seminars, of that magic atmosphere that takes shape when the analyst has delivered her or his interpretation (for instance, the interpretation of a painting by Rembrandt) and all the attendees are left with the image, the method, and the proposed reading, implicitly challenged by the speaker her- or himself to further improve that which has been proposed. Typically, at this stage two different kinds of questions are raised: on the one hand, someone points at an element of the painting that has not been taken into account by the meta-linguistic translation. Objections of this kind are interesting, but somehow fail to work as devices of corroboration. That happens, on the contrary (second kind), when some of the attendees not only point out at a neglected element in the painting and its meta-linguistic description: in addition, they are also able to reformulate the proposed interpretation in a way that encompasses all the elements that such interpretation takes into account, plus the neglected one. Confronted with remarks of this kind, the analyst will have to rejoice, since cooperative interpretive work has actually achieved more than a single analyst, albeit a genial one, could. In the third method, indeed, the community of scholars sharing the same method is more important than the single scholars, and it is actually more important than the method itself. What matters, indeed, is sharing, meaning that familiarity with the same meta-language, albeit at different levels of precision or sophistication, allows the analytical work to become a liturgy of interpretation, which etymologically means "work of the people". This liturgy will be guided by an expert, but necessarily followed by a series of practitioners of the analysis that share the same methodology and can advance each other's perception of the text. That is the way in which the Greimassian seminar in Paris, or Eco's seminars in Bologna and Urbino, or Lotman's seminar in Tartu worked in their halcyon days.

But why, then—given that focusing on the intentio operis allows one to create an inter-subjective arena where to debate, through the lenses of a meta-language, about the meaning of a text and, more generally, about interpretations—should the analyst be seized by that bittersweet melancholy that, as has been described at the beginning of this paper, subjugates her or him when the suspect arises that this laboratory of interpretation is actually an empty one, both because it does not resonate with the practices of interpretation that take place in the 'real world', outside of the academe, and because students and colleagues themselves seem to just pay lip service to the ritual of interpretation, turned into a sort of cold liturgy to which no one any longer attach any faith? One could not answer this question without accepting the saddening hypothesis that, in interpretation, it is not the method that creates the community, but the community that creates the method. That does not mean that some methodological proposals might not be more articulated and sophisticated than others. Greimas's proposal of a methodology for textual analysis was certainly more complex and rich than any method proposed by Roland Barthes, for instance. That, however, is not the point. The force of a method, like the force of a language, does not stem uniquely from the internal structure of the method, or from the internal structure of the meta-language, but from the fact that a community speaks such languages. Italian can work as a language not because it is phonetically, syntactically, or semantically coherent, but because a community of speakers believe and interiorise the belief of such coherence. Similarly, Greimas's method is effective not only per se, but because a community takes shape around this methodological proposal, and is willing to sacrifice interpretive idiosyncrasies in order to create a common arena for discussing interpretations.

That is the real cause for the interpretive melancholia that, nowadays, strikes many academics, especially those who lived in cultural epochs and contexts where, on the contrary, a feeling of the existence and the activity of an interpretive community would be strong. Looking outside of the classroom, what is it that the professional analyst sees? A chaotic cacophony of narcissistic voices, each boasting itself not only as a voice but also as the ultimate method that everyone should adopt; the transformation of relativism into the absence of dialogue; the conflation of the subjective method of interpretation with the inter-subjective one, a conflation wherein each interpreter claims to own and dominate the entire arena of interpretation; the total inability to embrace a principle of humbleness, according to which one's interpretive identity is at least partially sacrificed in the name of sharing a common space for discussion. In this interpretive semiosphere, it is natural that conspiracy theories, trolling, cynicism, and sectarianism thrive. Nobody wants to yield to a common method of interpretation for a fundamental mistaken is made, consisting in the inability to distinguish between accepting a common space in which to debate over interpretations, and accepting a common interpretation. The two options are completely different. A society in which all are forced to accept a single view is a dictatorial one. A society in which nobody accepts a space where a single view might arise, however, is an anarchist one. The two models are dangerous because the two models entail the possibility of violence, meant as agency that resolves interpretive conflicts in non-symbolical ways, not through persuasion but through force.

The melancholy of the interpreter arises even more when she or he sees that also in the purified atmosphere of the laboratory of interpretation, in the classroom, for instance, or in the conference hall, the pernicious cynicism that pervades society contaminates the instruments, disrupting any feeling of an interpretive community also among colleagues, or between professors and students.

4.8 Conclusion: Reasonable Manners in Interpretation

Is there any antidote to such melancholy? Again, a viable solution seems to lies in what Umberto Eco has underlined all through his career, that is, the essential difference between rationality and reasonableness (Eco 1992). What is reasonableness, also etymologically? It is the potentiality of rationality. Something that is reasonable is not rational per se, but can be recognized as such within the continuous exchange and negotiation of a community devotedly sharing a public discourse. Claiming that an interpretation is the only rational one, especially in the field of humanities, dangerously resembles an act of authoritarianism. There is no single way to interpret the meaning of a novel, or that of a painting, or that of a film, and fortunately there will never be. Language is slippery and mutable, not a solid but a liquid substance, or even an airy one in the case of poetry or similar discursive formations. One should strongly object, however, that claiming that there is no single rational interpretation of a cultural fact is tantamount to claiming that there is no single reasonable interpretation of it. When I affirm the rationality of my interpretation, I impose my view, mostly by authority if not even by force. But when I affirm the reasonableness of my interpretation, I do not impose it; on the contrary, I offer it to a community to which I myself subscribe, by curtailing my cynicism, my narcissism, my envy, and by sharing a methodology that, although it might not be perfect, although it might be perfected, is nevertheless able to create a space of inter-subjective recognition, where the correctness of an interpretation might always be recognized.

Admitting and preserving this difference is fundamental not only to protect an arena for the discussion and corroboration of interpretive hypotheses but also because such arena is essential also for their falsification. Claiming that reasonable interpretations do not exist, indeed, is tantamount not only to thwarting any possibility of ranking them positively, but also to undermining any opportunity to ranking them negatively. In a society that is unable to humbly turn itself into a structured community of interpreters, one can neither hope that a new interpretation of Shakespeare will be proven as more encompassing and satisfying than any previous one, and that a new conspiratorial revisionism will be disproven as less reasonable and, therefore, less tenable than any preceding interpretation of history or society.

After all, what is needed for a society to work as an interpretive community is not a perfect method of interpretation. It is reasonable manners.

References

Aumont, Jacques. 1990. *L'image*. Paris: Armand Colin. English translation by Claire Pajackowska. 1997. *The image*. London: British Film Institute.

Barnstone, Willis. 1993. *The Poetics of translation: History, theory, practice*. New Haven, CT: Yale University Press.

Calabrese, Omar. 2000. Lo strano caso dell'equivalenza imperfetta. (Modeste osservazioni sulla produzione intersemiotica). *Versus* 85–87: 101–120.

Danesi, Marcel. 2002. *The puzzle instinct: The meaning of puzzles in human life*. Bloomington, IN: Indiana University Press.

Eco, Umberto. 1976. *A theory of semiotics*. Bloomington, IN: Indiana University Press.

Eco, Umberto. 1990. *The limits of interpretation*. Bloomington, IN: Indiana University Press.

Eco, Umberto. 1992. *Interpretation and overinterpretation*; with Richard Rorty, Jonathan Culler, and Christine Brooke-Rose, ed. Stefan Collini. Cambridge, UK: Cambridge University Press.

Fawcett, Antoinette, Karla L. Guadarrama García, and Rebecca Hyde Parker (eds.). 2010. *Translation: Theory and practice in dialogue*. London and New York: Continuum.

Genette, Gérard. 1987. *Seuils*. Paris: Seuil. English translation by Jane E. Lewin; foreword by Richard Macksey. 1997. *Paratexts: Thresholds of Interpretation*. Cambridge and New York: Cambridge University Press.

Greimas, Algirdas Julien. 1976. *Maupassant: La sémiotique du texte: Exercices pratiques*. Paris: Éditions du Seuil; English translation by Paul Perron. 1988. *Maupassant: The Semiotics of Text: Practical Exercises*. Amsterdam and Philadelphia: J. Benjamins Pub. Co.

Hjelmslev, Louis T., and Hans Jørgen Uldall. 1957. *Outline of glossematics: A study in the methodology of the humanities with special reference to linguistics*. Copenhagen: Nordisk Sprog- og Kulturforlag.

Kreuz, Roger J., and Mary Sue MacNealy (eds.). 1996. *Empirical approaches to literature and aesthetics*. Norwood, NJ: Ablex Pub.

Leone, Massimo. 2013. The semiotics of fundamentalist authoriality. *International Journal for the Semiotics of Law—Revue Internationale de Sémiotique Juridique* 26 (1): 227–239.

Leone, Massimo. 31 December 2016. Metafisica del design: il senso degli oggetti in De Chirico, Kiarostami, Ozu. *E/C*, Online Journal of the Italian Association for Semiotic Studies: 1–37.

Maitland, Sarah. 2017. *What is cultural translation?*. London, UK and New York, NY: Bloomsbury Academic.

Oblinger, Diana G. (ed.). 2006. *Learning spaces*. Washington, DC: Educause.

Popper, Karl. 1935. *Logik der Forschung: zur Erkenntnistheorie der Modernen Naturwissenschaft*. Vienna: Springer. English translation by Id. *The logic of scientific discovery*. London: Hutchinson, 1959.

Reason, James. 2013. *A life in error: From little slips to big disasters*. Farnham, Surrey, England: Ashgate.

Struck, Peter T. 2016. *Divination and human nature: A cognitive history of intuition in classical antiquity*. Princeton, NJ: Princeton University Press.

Zhang, Jiang. 2016. The dogmatic character of imposed interpretation. *Social Sciences in China* 37 (3): 132–147.

Chapter 5
The Game of Reflection and the Power Over People. A Semiotic Approach to Communication

Dumitru Borțun

Abstract Ogden and Richards emphasized that in an act of communication the message does not exist before being coded; in addition, coding is a process of creation: *the message self-generates in the communication process itself.* As such, the term "receptor" should be replaced by the term "reader"—in other words, the "universal receptor" breaks into a multitude of readers, whose readings are culturally pre-determined. The significance of a sign is not given beforehand, it is born following the encounter between the message and the cultural loading with which the reader welcomes the message. As *homo significans,* we relate not to objects, but to "interpretants" (Peirce). *For man, the world is a universe of interpretants.* But this renders manipulation possible through partial truths or even fake news, i.e. through the *plausible denaturation of reality*. The acceptance of a phrase as true is not related to its relation with reality, but rather to its relation with the reader's cultural loading. This paper describes the mechanism of semiosis that makes possible the exercise of power to have people over other people, through the management of their cultural loading.

5.1 Separation Between Reality and Its Reflection

When naming the science of signs "semiology", Ferninand de Saussure fancied that he had just laid the foundation of a new science which would study *semiosis*, i.e. an alternative way in which man could assimilate and appropriate the real, other than through knowledge. *Homo cogitans* was just joined by *Homo semnificans.*

With Saussure, we grasp the idea that language is a structure of thinking which would exist regardless of linguistic forms, since there is a difference between language (*la langue*) and speech (*la parole*) (Ducrot and Todorov 1972: 44).

D. Borțun (✉)
National University of Political Studies and Public Administration,
Bucharest, Romania
e-mail: dumitru.bortun@comunicare.ro

A. Olteanu et al. (eds.), *Meanings & Co.*, Numanities - Arts and Humanities in Progress 6, https://doi.org/10.1007/978-3-319-91986-7_5

According to Oswald Ducrot "the Saussurian thesis proves that there is a fundamental linguistic arbitrariness, different from the arbitrariness of each isolated sign. Thought exists before language, as an amorphous mass, as a nebulosity, devoid of privileged structures" (Ducrot and Todorov 1972: 30).

Saussure was the advocate of the "scientific" approach to language, i.e. of the approach subordinated to the logical-mathematical ideal of intelligibility. He stated that: "It must be conceded that the theoretically ideal form a science should take is not always the form imposed upon it by practical necessities. In linguistics, practical necessities are more demanding than in any other subject. To some extent, the confusion which at present reigns in linguistic research is due to them. [...] The ideal program would be for each scholar to concentrate either on the synchronic or diachronic research and include as much as possible of the material falling within his chosen field; but it is difficult to achieve a scientific understanding of widely different languages" (Ducrot and Todorov 1972: 112–113). Unfortunately, from the logical analysis of language, be it structuralist (*avant la lettre*), we discover nothing important neither about the "life of language", nor about a better use of language in our lives. The finding that language is a system of signs is awe-inspiring but, in spite of its awesomeness, it fails to take us beyond language, in the world of signification.

The main contribution to the birth of semiotics as an autonomous science belongs to Charles Sanders Peirce who defined the sign as follows: "a Sign, or *Representant*, is a First which stands in such a genuine triadic relation to a Second, called its *Object*, as to be capable of determining a Third, called its *Interpretant*, to assume the same triadic relation to its Object in which it stands itself to the same Object" (Ducrot and Todorov 1972: 41). For Peirce (1931: 5.484), *semiosis* is a relation between three terms (sign, object and interpretant): something that triggers the chain process, the object existing in reality and the effect that the sign produces on the interpretant. The interpretant is also a sign for another Interpretant—and so on, up to perfect Signs (for instance, the relations between the word and the terms that define it in the dictionary: synonymy or paraphrasing).

As such, the human world is not a world of objects, as it is for the non-human animals; as opposed to the latter, which interact with things and phenomena, man interacts with their meanings. As *homo semnificans*, the human relates to interpretants. For us, humans, the world is a universe of interpretants, rather than a collection of things with traits between which there are different relations. If it were so, we would be unable to perceive the sunset or the sunrise.

In 1930, the Romanian poet Ion Barbu (1895–1961), renowned also as mathematician, said that poetry was a "purer second game"—purer than the "wild herds", i.e. the reality outside the subject (*out there*).[1] If we take the model of understanding

[1]Excerpt from the poem entitled *A Hidden Game*: "From time, abstracted the depth of this peaceful crest, Gone through the mirror into redeemed azure. Engraving on the sinking flocks of rustic fest. Out of the water groups, a second game, more pure. Latent Nadir! The poet elevates summation. Of spread out harps you lose in a reverted flight. And painfully distils a song: hidden, as only sea's cremation. Sways its Medusas under the greenish bells of light" (tr. Liviu Georgescu, http://www.poezia.3x.ro/BARBUi5.htm).

the world proposed by Peirce seriously, we have to accept that not poetry, but *perceived reality* itself is "a second game", a game of reflection, and its poetic representation is, in the best scenario, only a third game. As Peirce would put it, we, individuals, enter into a first contact with signs that stand for objects and which produce in our minds other signs, called interpretants (game of reflection): "A sign is something which stands to somebody for something in some respect or capacity. It addresses somebody, that is, creates in the mind of that person an equivalent sign, or perhaps a more developed sign. That sign which it creates I call the *interpretant* of the first sign. The *sign* stands for something, its object" (Peirce 1931: 2.228).

In *Introduction to Communication Sciences,* John Fiske explains the production process of the *interpretant* which he equals with the "meaning of the message": in the semiotic approach to communication, the message is a construction of signs which, through interacting with the receivers, produce[s] meanings (1990: 3) (*the meaning of the message*). In this approach, the focus is not so much on communication as process, but rather on communication as the generator of meaning. The sender (message transmitter) loses his importance. The focus is directed towards the "text" and the way it is "read". "The reading" is the process of discovering the meaning that emerges when the "reader" interacts or *negotiates* with the "text". As regards the signs and the codes of the message, the "reader" filters the message through the strainer of the *cultural loading*. The more we share the same codes and the same sign system, the closer the two significances attributed to the message (Fiske 1990: 40). In Fig. 5.1, we have tried to illustrate the mechanism of meaning production, as negotiation, i.e. as interaction between a sign and an *active subject* ("active" in a Kantian understanding).

The subject can read the message by virtue of its internal structures: logical, linguistic, psychological, axiological, and symbolic (=*Cultural Loading*)

The cultural loading consists in linguistic, logical, psychological and symbolic structures which get out to meeting the message and "negotiating" with the structures similar to it. We saw that in the semiotic approach of communication, this is a *generation* of significance through messages, generation carried out either by the one who codes or by the one who decodes the message. Meaning is not a static, absolute, clearly defined concept in the message; it is an active process. In order to characterise this process, semioticians use verbs such as *to create, to generate* or *to negotiate.* As such, the content of a message is not a mere transfer of something under a certain form or due to a certain form, as in the common sense vision,

Fig. 5.1 Production of meaning as interaction

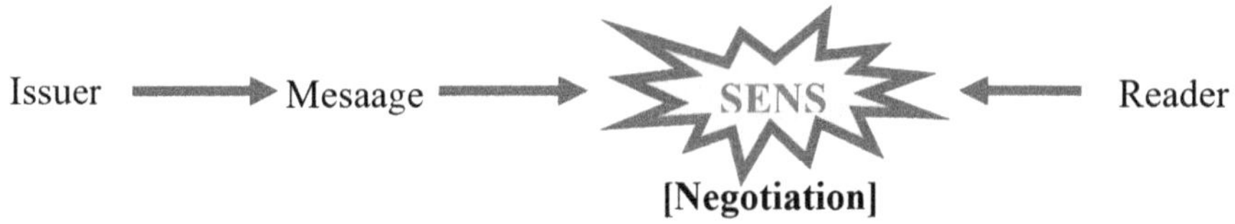

Fig. 5.2 Genesis of the message as negotiation

which Richards ironically called "the vulgar packaging theory of communication" (Ogden and Richards 1923 [1949]; *apud* Fiske 1990: 26). Richards used this phrase to blow off the steam against some approaches such as Shannon-Weaver, which imply that there is a core of the message, prior and independent in relation to form. It is coded in a language (i.e. it is "packed" as a postal package) and then it is sent through a channel (by mail). In order to consume the message, the receptor only has to decode it (to "unpack" the package). In Richards's vision, the message does not exist before it is coded (expressed in an articulate language), and the coding of the message is a creative process. During the transmission of the message and on that occasion, the subject creates or modifies meanings: the message *self-generates in the communication process itself.* This is what Fig. 5.2 illustrates.

This is why Richards compared the most popular scheme of communication (Issuer → Message + Channel → Receiver) with a vulgar packaging theory. We may call it vulgar, but also naïve, because this is the way common sense assumes that the communication process unfolds: an addresser buys a gift from a shop (a message), packs it up (codes it), sends it by mail (through a channel) and the addressee unpacks it (decodes it) and uses it somehow (consumes it). Unfortunately, we may say, there is no messages shop, as there is no such thing as the world of ideas depicted by Plato. There is no such place where we could make message supplies as common sense may presume. Its main illusion related to inter-human communication is to reduce the message to information and communication to an exchange of information. A good book contains information, but its message is born in our mind, as a final product of reading or it is not born at all.

5.2 Relativity of the Interpretants and Cultural Relativism

Therefore, the message is not something prior to the communication process, which would exist independently from the Issuer-Receptor interaction and sent by the Issuer toward the Receptor, but rather an element within a structured relation which includes, among others, the external reality, the producer of the message and the reader of the message (the creator of significance). The production and reading of the text are regarded as parallel (if not even identical); within these processes, the relation is structured so that they occupy the same place. The concomitance between communication and generation of meaning, that we deem to be the very essence of the semiotic approach to communication, is of utmost importance for the understanding manipulation. What does "target" mean if not a group that shares the same cultural load?

As we have shown, from a semiotic perspective, the acceptance or the rejection of a message is not connected to its relation to reality, but to its relation to the cultural loading of the reader. By activating this cultural loading, the addressee is upgraded from the status of *receptor* to the one of *reader*. As such, reading reality becomes a product of the negotiation between this cultural loading and the structure of the message. According to John Fiske, from the terms used by semioticians, negotiation is, probably, the most useful since it "implies a to-and-fro, the-give and-take between person and message" (Fiske 1990: 46).

Judging through semiotic lenses, observing the constant, bi-univocal interactions between the message "producer" and the reference system, between these and the "reader", we deal with the cultural determinism of communication. The link between the semiotic approach and *intercultural* communication becomes obvious once we understand culture in the paradigm of cultural anthropology—for instance, as defined by E. B. Tylor, T. Parsons or Chombart de Lauwe. In the introductory study to the volume *Images de la culture*, entitled „Systemes de valeurs et aspirations culturelles", Paul-Henry Chombart de Lauwe classifies approaches to culture as follows: (1) culture as a development of the individual in society, (2) cultures of some societies or particular social environments and (3) the issue of the development of a universal culture. It goes without saying that of the three approaches, the only that does not presuppose a prior assessment and which does not necessarily lead to the hierarchization of cultures (societies, groups or individuals) is the second one. It shall, in fact, be the privileged referential of this paper, since it best corresponds to its objectives. Insisting upon the role of infrastructure generating desires and systems of value, Chombart de Lauwe considers that "a culture is marked by a range of models, guiding images and representations affecting the behavior, work, roles and social relations of the members of a certain society". He calls upon the equal importance of *techniques, space organization, production and work or consumption* (Chombart de Lauwe 1970).

Without knowing what the tacit and the explicit knowledge mean, pre-determining the code in which a message is assimilated (according to Fig. 5.3), we cannot grasp the image created in the mind of each addressee. In the example imagined by us and illustrated in Fig. 5.3, a Signifier ("student in a hurry") can generate several Signified, depending on the tacit knowledge contained in the cultural paradigm of each of the receptors: a student, a teacher and a college registrar. In Peirce's terms, a Sign (first sign) can generate, in the mind of different individuals, different Interpretants (different secondary signs). The path from Sign to Interpretant passes through the knowledge-language pyramid, similar to the way in which the rays of light pass through a prism.

This is the reason why, in communication campaigns, a prerequisite for success in the behavioral change of a given public is the knowledge of the cultural paradigm that defines the targeted audience, knowledge that can be obtained by mixing the theoretical approach with the sociological research, conducted by a specialized

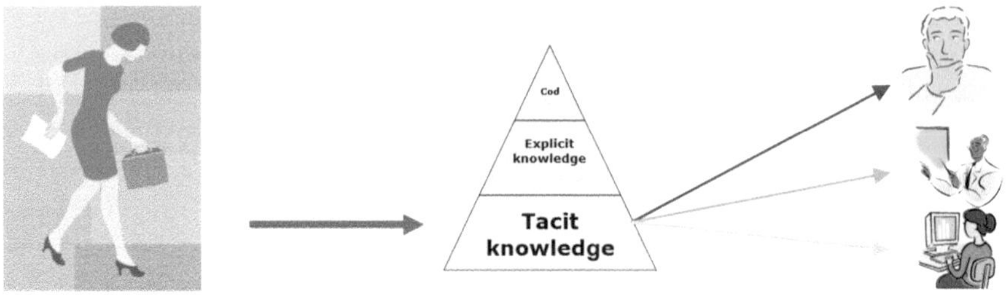

Fig. 5.3 The filter enforced by the cultural paradigm

institute. As a last resort, we are dealing with the knowledge of the value-attitude pairs (Linton 1945), because they are directly reflected in the behavior: the change of a behavior presupposes the change of attitudes! This, in the 20th century, becomes simpler and simpler: *the efficient management of interpretants*. This is why one of the definitions of public relations, all too when dealing with public image management, with branding and rebranding or reputation management, is the "management of social representations"—aspect that we have developed elsewhere (Borţun 2012: 118).

5.3 Consequences Related to the Theory and Practice of Manipulation

Manipulation, as form of gaining control over individuals, stems from the possibility of guiding the reading of reality by structuring the message in correlation with the information held by the issuer regarding the addressee's cultural load.

As shown elsewhere (Borţun 2002), the negotiation process gives rise to Significance—in fact, to the very message itself, as there is no message without significance, as there is no sign without significance (if a sign is transplanted into a culture in which it receives no significance, it shall not function as sign in this culture). When the receptor does not hold all the structures necessary for reading, he "perceives" but is ignorant of what precisely he perceives (the so-called state of perplexity); when the receptor holds none of such structures, the message that is sent toward him shall not encounter any significance-generating structure, it shall be as real as a ray of light encountering no object reflecting it and which gets lost in the "cosmic darkness". This is why, in the elaboration of an efficient and effective communication, the research phase is mandatory in view of knowing the cultural loading of the target audience, for the messages to be so conceived that their path crosses the significant structures of this loading.

The differences between the various cultural loadings typically reproduce the differences between cultures, between sub-cultures (including the so-called "professional cultures", "local cultures" or "organizational cultures"). From one culture to another, the same sign can produce different interpretants—leading to the issue of intercultural communication, of multiculturalism and cultural relativism.

Cultural relativism is a natural result of the bankruptcy recorded by *logocentrism*, a vision for which verbal language is the uncrowned king of the two neighboring kingdoms, Knowledge and Communication: of Knowledge because it is the main tool of the much desired absolute objective knowledge (see the Vienna Circle and young Wittgenstein's programs); of Communication because it is a "mandatory passing point" for all the other languages (artistic, musical, cinematographic etc.): in order to become efficient communication tools, they must be translated into verbal language—see the armies of critics of art, music critics or movie critics, whose mission is the universalization of messages, by re-coding them, enabling their readers to understand them "correctly", regardless of their artistic, musical or cinematographic culture (see also Eriksen Stjernfelt 2010).

The first consequence of this "linguistic monotheism" was the *intellectual illusion* (Iliescu 1989: 124–126), expressed in two methodological suppositions: *the prevalence of intension over extension* (in other words, the logical provisions are primordial in relation to the use of expressions within a form of life)[2]; *the autonomy of semantics in relation to pragmatics* (from a methodological standpoint, it meant the marginalization of the interest in everything related to the psycho-social use of language, of the "life of language"). It is one of the causes through which some authors explain the rejection of globalization in numerous forms of life: it is perceived as the *Westernization of the world* (Matei 2016: 67–68).

With Wittgenstein, we understood that these suppositions are illusory, that meaning is not an *entity* (psychological or logical), but rather an *activity*; it is neither invisible, nor is it attached to the expression, but a way of operating with signals—case in which thinking cannot be anything else but speech (Wittgenstein 1980). In a blink of an eye, the construction of meaning began to be monopolized by various instances: firstly by the propaganda apparatus of totalitarian regimes (Stalinism, Nazism etc.) and, secondly, by the advertising industry. The analysis conducted by Francoise Thom regarding the wooden language as instrument of exercising power (Thom 1987) is the description of a monopolization mechanism of semiosis and of enforcing, at social level, some "official interpretants". As regards the non-Stalinist regime in Romania, the unique party exercised its control over language according to the three rules identified by Lavinia Betea: the rule of selection, the rule of constraint and the rule of words' hierarchization. The latter is based on the axiom that individuals' political opinions depend on the vocabulary they use (Betea 2000: 29). Language also contributed to the manipulation of social memory, by appealing to lies: *fake news* about the adversary, *fake news* in the information sphere and *fake news* through which reality was re-created (Betea 2000: 16). When Peirce defined the interpretant as "an equivalent sign, or perhaps a

[2]Which equals to the prevalence of the essence over the existence, because, as noted by Adrian-Paul Iliescu, "the idea of *essence* returns in the modern philosophy disguised under the name of *intension*" (Iliescu 1989: 124).

more developed sign that another sign created in someone's mind" (Peirce 1931: 2.228), he could not have imagined how "developed", even "*falsified*" can this sign be in relation to the first sign.

With reference to the *advertising* industry, the monopolization of the construction of meaning began to be emerged in correlation with the progressive marketization of some domains of society, including communication. This does not refer only to the political discourse, where marketization led to the de-politization of the political discourse (Borţun 2011), but also to the types of discourse perceived up until recently as immune to subjective and biased interpretations, such as the scientific discourse or the news discourse.[3]

The last frontier of marketing is the so-called "experiential marketing" which is grounded in the following axiom: "The experiential approach recognizes that the human being has both a cognitive side and an emotional side (Addis and Holbrook 2001). The concept of experiential marketing has been used for the first time in 1999 by Bernd Schmitt. In his opinion, experiential marketing does not refer to the quality, the advantages and the functions of the promoted product or service, but to the emotions that the consumer gets and the feelings that it activates following the consumption experience. Consumption experience is experience projected from the perspective of experiential marketing (Schmitt 1999: 3). Furthermore, Lee and col. defined the product of experiential marketing as being "a memorable idea or an experience deeply embedded in the consumer's mind" (Lee et al. 2011). According to another promoter of experiential marketing, it is "the process of identifying and satisfying customer needs and aspirations profitably, engaging them through two-way communications that bring brand personalities to life and add value to the target audience" (Smilansky 2009).

We cannot image a more precise description of the semiotic shift in marketing in the last two decades: the sign sent to the reader does not necessarily have to have several links to the "qualities, advantages and functions" of the object for which it stands, but it has to be suited to the "needs and aspirations" of the reader to whom it has to convey certain emotions and in whom it activates certain feelings so that the interpretant born in his mind is either a "memorable idea" or "an experience deeply rooted in the consumer's mind". The main purpose of the birth of interpretants is the development and the maintenance of an emotional attachment of the consumers toward the respective brand.

When Richard Rorty indicated the re-description as a means of changing perceptions and, consequently, reality itself (Rorty 1989), the dice had already been cast, semiosis had already been taken under the control of governments and corporations and toward the post-truth epoch there was only a step. To use a cliché, we could say that the evil had already been done. What exactly is this evil? The evil consists in the destruction of the harmony of semiosis, i.e. of the cooperation

[3]Related to the re-orientation of such types of discourses from facts toward cultural loading of the reader, through the autonomization of the interpretant in relation to the object, see Bruno (1991), for the scientific discourse, or Borţun (2014), for the news discourse.

between object, sign and interpretant[4]—more precisely, in the *autonomization of the interpretant in relation to the object, through arbitrary manipulation of the sign*. In the "post-moralist epoch", marked by the "marketing of values" and through the "individualistic invention of one's own person" (Lipovetsky 1996: 86–89), *the use of the sign tailors itself less and less to the object and more and more to the cultural loading of the target audiences*. Only one step separated us from the de-valorization of facts and to the authority crisis of the idea of objective reality, proper to the post-truth epoch. And this step has been, in fact, taken, as we shall further see.

I believe it is fortuitous that this process emerged at the initiative of anti-religious regimes, hostile to the tradition of semiotics as developed in medieval Christianity: Stalinism and Nazism. From the perspective of Christian semiotics,[5] between the object and the sign for which it stands there is a relation that leaves no space for fake news, that does not allow the deliberate denaturation of reality.

Almost one decade ago, my former student Alin Olteanu, obtained the Bachelor's Degree with a thesis on Saint Augustin's semiotics (Olteanu 2009). He began with two confusing statements of Umberto Eco: semiotics is the discipline "concerned with everything that can be taken as a sign" and the sign is "everything that can be taken as significantly substituting for something else" (Eco 1976: 7). This *something else*—said Alin Olteanu—can exist or not, from a semiotic point of view the existence of the signified object is irrelevant: "In this way, *something* can be used to tell the truth to the extent that something can be used to lie. *Something* that cannot be used to lie cannot be used for anything, not even for telling the truth" (Olteanu 2009).

Challenged by the epistemological ambiguity on which semiotics is substantiated, Olteanu traced back the history of this discipline in order to detect in which moment of its evolution the possibility of deceit and self-deceit appeared, when exactly the *plausible denaturation of reality* became possible. He found in Saint Augustin an ideal model of communication, in which signs (words) become useless, and the interpretant is the object itself, which "speaks" by itself.[6] Answering to the question "In what langue did God Speak with Adam in Paradise?", Saint Augustine initially preaches the official version of the Hebrew language as the "human language of the origins" (*De civitate dei*, XVI: 11, 1). Then, he says that such a communication, of divine nature, does no longer presuppose a means resembling a human language (*De vera religione,* XXXIII: 61). As such, within an ideal model of knowledge and communication, authentic knowledge consists in the reading of the sigs revealing to the reader the Truth, *id est* Divinity. Lying is

[4]For Charles S. Peirce, semiosis is "an action, or influence, which is, or involves, a cooperation of three subjects, such as a sign, its object, and its interpretant" (*apud* Eco 1976: 32).

[5]See Andreopoulos (2006), Ciapalo (1997), Deely (2001) or Marmo (2010).

[6]In Saussurian language, the signifier becomes useless and the signified coincides with the object, since semiosis does no longer imply any agency.

impossible, wanting to conceal the truth is a sin (in the Christian perspective) and the proof of the sin is fake knowledge, i.e. the erroneous interpretation.[7]

In Augustine's vision, everything seems to be read for the purpose of knowing the Truth, even for the purpose of lying: "As such, the corporeal form, since it is devoid of will, does not lie. Because, even if it is not considered to be what it is not, it is still not deceitful" (Augustin 2007: 135). Through this statement, Saint Augustine showed that any significant can lead to the correct significance, the single problem being interpretation. After all, humans can neither linguistically reveal the Truth, but neither can they lie. How could man conceal the Truth? Olteanu draws the conclusion that for Saint Augustine, the self is the only one responsible for what he knows: good or evil, beautiful or ugly, true or false: "Everything is good, everything is beautiful, and everything is true. Human knowledge is subject to mistake, and the error is rooted in human language. Truth is there, in front of the spirit, of the senses and of the intellect, it just has to be read" (my translation, Olteanu 2009). Recently, Olteanu found interestingly similar arguments in other medieval scholars, such as Dionysius the Areopagite and Ibn Arabi (Olteanu 2018).

The illusion of unity between sign and interpretant (between signifier and signified, respectively) has been denounced by Wittgenstein who understood that, due to the improper use of language, the interpretant can emancipate from the sign up to total autonomy and, to the limit, it can go against the object.[8] From Stalin's "kulak" and Hitler's "rats" up to the "New Europe" of Donald Rumsefeld and the "iliberal democracy" of Viktor Orban,[9] we are dealing with a chain of "re-descriptions" (Rorty 1989) which impacted on reality, including on the reference objects. Among these consequences, the most important are the ethical ones.

Contrary to his friends, Carnap, Frege and Russell, Wittgenstein understood that the mere logical analysis of language cannot teach us anything relevant about how we should live. He wanted to see what is beyond language and he found out: Ethics.[10] The relation between technical analysis, perceived as immoral, and the ethical stakes of the author is reproduced by Wittgenstein in an almost identical manner as in Adam Smith. As the *Tractatus* received from the Anglo-Saxon critics a "logical" interpretation, unchallenged for 50 years, until Janik and Toulmin (1973), Smith's work *The Wealth of Nations* has been read, commented upon and

---

[7]In his vision, the main source of error in the semiosis process is language, conviction that we shall later find at Hilbert, la Carnap and Frege or young Wittgenstein, who will attempt to regulate language up to the point in which it leads to no interpretation errors of reality.

[8]Gulag, Holocaust and other disasters of the 20th century have been accepted by a significant number of individuals and, because of the interpretants' swaying away, in the diabolic labs of Stalinist and Nazist propaganda.

[9]The term has been launched by Fareed Rafiq Zakaria, in 1997, being used by political scientists to describe the current political regimes in Poland, Russia and Hungary (Pârvulescu 2017).

[10]According to Allan S. Janik and Stephen S. Toulmin, Wittgenstein's interest in language has had, from the very beginning, an ethical stake, but he because progressively aware of it. Since it was not clear from the start, he asked confusing questions to his friends from the Vienna Circle, and, as a result, it is said that they sent his to Cambridge, hoping that Russell would understand that their young fried wanted to find out (Janik and Toulmin 1973).

taught for over two centuries, just as a political economics paper. The author was more interested in the exchange relations because they tell us something significant about our lives,[11] rather passed unnoticed.

In the modern epoch, the ideal of an absolute objective knowledge, of a unique and infallible Truth, embraced by all rational individuals has been taken over by science, propagating either inertly (at the level of common knowledge) or methodologically (at the level of scientific knowledge), under the form of Hilbertian ideal of total axiomatization, or programmatically (at the level of philosophy and epistemology), under the form of logical positivism and logical semantics. On a scientific, epistemological and philosophical level, "disenchantment" took place from the inside, along with Gődel's theory (1931). But on the level of common knowledge, faith in the possibility of an infallible truth has been cleaved from outside knowledge: in the 20th century, an already secularized century, the two World Wars, waged in the name of some truths, the Gulag and the Holocaust, part of a monstrous mechanism whereby the truth of some individuals was enforced upon the others, were the final blow to the organic connection between the object and the interpretant—between object and significant, respectively. Men understood that signs (or significants) can be used in the name of the truth as well as for serving lies—and even worse, that certain interpretants can account for the abdication from humanity and may lead to a large-scale carnage, to the industrialization of death. They could already get a glimpse of the post-truth epoch.

5.4 The Post-truth World and the Battle of Interpretants

The term *post-truth* is defined by *Cambridge Dictionary* as follows: "men tend to accept arguments based on their emotions and beliefs, rather than ones based on facts". As such, it is conceived as a *de-valorization of facts*, which no longer represent a reference for accepting true judgements and rejecting false judgements, along with a *crisis of the idea of objective reality*, which no longer represents a truth-founding instance. As regards the loss of prestige of "facts", it has been announced by Nietzsche who noted in a notebook published posthumously that "there are no facts, just interpretations".[12] The idea has been taken over, with other arguments, by the post-positivist epistemologists among whom the most important

[11]The moralist Adam Smith, who had published, 17 years before, *The Theory of Moral Sentiments* (1759), was not interested in poverty as a strictly economic phenomenon, but rather in *misery*, which is material as well as moral; he was not interested in wealth, but in *prosperity*, as a form of civilization, as a society that allows its members to be moral: "a polite society".

[12]„ … nein, gerade Thatsachen giebt es nicht, nur Interpretationen" (Nietzsche, *Nachgelassene Fragmente 1885–1887, Krtitische Studienausgabe*, hg. von G. Colti und M. Montinari, Munchen, 1988, Bd. 12, p. 315 (*apud* Goian 2017: iii–iv). The idea has been translated, with other arguments, by the post-positivist epistemologists that Raymond Boudon called „anthropologists of science", among them, the most important ones are Kuhn and Feyerabend, whom he reproaches the thesis according to which "facts are that scholars consider to be facts" (Boudon 593).

are Kuhn and Feyerabend. The two were downplayed by Raymond Boudon who called them “anthropologists of science”, reproaching them the thesis according to which “facts are that the scholars consider to be facts” (Boudon 1997: 593). According to Kuhn’s theory, two researchers embracing different disciplinary paradigms may look in the same direction and from the same point and perceive different things (Kuhn 1970: 92–93). This happens because of the tacit knowledge contained in the two paradigms, which interposes on the stimulus-perception circuit (phenomenon that we illustrated in Fig. 5.3). According to Kuhn, the famous urge “Let’s go to the facts!” no longer triggers waves of enthusiasm.

In connection to the crisis of the idea of objective reality, we may say that post-truth reflects a more serious crisis: not only of the *idea*, but of the *objective reality* itself. It is the concern that also tormented Raymond Boudon when he was criticizing “the anthropologists of science” because they reached the conclusion that “the notion according to which there is a reality outside us (*out there*) is a simple illusion” (Boudon 1997: 593). However, for Boudon this was more of an epistemological concern.

We are currently witnessing an actual attack against objective reality, but not from the perspective of reason, critical thinking, scientific knowledge, technique and technology, as it happened in the modern era, under the Illuminist imperative of the transformation of reality by humans according to their needs and ideals. Bacon’s slogan, “Science is power”, which we may consider one of the mottos of modernity, falls hazardously. Currently power no longer belongs to the ones holding true knowledge about objective reality, but to those who impose their own *interpretation* on this reality, generating the emergence of the so-called “information war” which we shall tackle in the following section. Quoting the poet Ion Barbu, the first reality is replaced by the “game of reflection”, which is perceived as reality; in the terms proposed herein, *Knowledge* is dethroned by *Semiosis*. At present, objective reality is attacked by the drives of irrationality, of *wishful thinking*, of biases and illusions, of phantasms taken as reality, of opinions taken as knowledge, of utopias considered feasible. According to an inspiring phrase of Peter Dale Scott, “post-truth may replace reality, may undermine it, may destroy it” (Goian 2017: i). How can it destroy it? An illustrative example is Merton’s “the creative prophecy” (Merton 1998).

In 1942 Robert Merton drew attention over a special prediction: albeit false, it becomes true if a sufficient number of individuals believe in it. In education, the pessimistic prognosis related to the evolution of a child may lead to his/her failure; in economics, the false prediction regarding an imminent inflation may prompt inflation, the false prediction concerning the scarcity of some products may trigger the increase of the cost thereof and, implicitly, poverty, whereas the functioning of a stock exchange often entails creative prophecy—especially under the form of the self-fulfilling prophecy.[13] In politics, the prediction of journalists regarding the nomination of a party’s candidate for presidential elections may lead to its

[13] A prophecy can not only be self-fulfilling, but also self-destructive when its very formulation may lead to the non-fulfillment of the predicted events, although upon its formulation it was true.

nomination and whereas the prediction made in the opinion polls regarding the winner of the elections may generated, under certain circumstances, his/her success. Starting with 1947, in sociology experts have been using the so-called "theory of Thomas", according to which *a false fact, if perceived as real, becomes real in their consequences* (the case of a bank which goes bankrupt following the launching of a false rumor that it went bankrupt).

The term *post-truth* has been first used in 1992 by Steve Tesich, who argued, in an essay published in *Nation Magazine*: "we, as a free people, have freely decided that we want to live in some post-truth world". In other words, people accepted the idea that they live in a world of interpretants after they believed, over centuries, that they were living in a world of objects characterized by *traits* and which are *related* between them, a world reflected as in a mirror by language, by means of *nous, adjectives and verbs,* respectively.[14]

For the great majority of people, the bankruptcy of the *language-mirror* concept, firstly coined by constructivist epistemology, equaled with the bankruptcy of the *objective reality*, of which we can no longer be sure, which we are no longer bound to take into account. Therefore, truth ceases to be a legitimation value in the public space—process depicted with a surgical accuracy by Rorty (1989). We witness a process of "democratization of truth", in which the difference between knowledge and opinion is no longer valid, nor is the difference between amateur and professional, between profane and initiated, between neophyte and professional—process which is plane to see on the social networks where we assist to a fascinating show: *the battle of the interpretants*. A battle waged in the name of the truth, but of an unprovable and, at the same time, indisputable truth.

5.5 The Information War as Semiotic War

The expression "battle of the interpretants" oversimplifies the reality I refer to. In fact, interpretants cannot be autonomous in relation to the objects and the signs that stands for them; this is why a more rigorous expression could be "battle between semioses". Charles Sanders Peirce perceived the process of semiosis as follows: "Semiosis is an action, or influence, which is, or involves, a cooperation of three subjects, such as a sign, its object, and its interpretant, this tri-relative influence not being in any way resolvable into actions between pairs" (Peirce 1931: 5.484). The disinformation strategy by means of *fake news* has intrinsic limits: it does not manipulate, but misinforms and may contribute to manipulation. As we can see in Fig. 5.2, information is just the "message" that the "reader" encounters, its reception being in fact an extremely complex activity, that we call "reading". A product of manipulation can only be

[14]It is more an explicit recognition of the fact that people do not need truth, the performance of a consensus unfolded over this. The tacit avoidance of the truth dates long back and, if we were to believe Jean-Francois Revel, the marginalization of this value is a constant aspect of human history (Revel 2007: 231–235).

considered the *result of the reading*, but only when and because a conscious subject deliberately influenced the structures (the cultural loading).

In connection to the difference between disinformation and manipulation, experts in information argue that: "We need [...], instead of a simplistic and sociologically uneducated causality, a two-step-thinking. The first level is the research of the degree of social trust in the institutions of a state or of a more ample entity (EU, NATO etc.). When it is solid, the odds of success of the destabilization actions from an external enemy by means of a media attack are low. When the respective society already faces a crisis, when the wounds are bleeding, when the level of trust in the institutions (in the "System") collapses—then the external intervention is undoubtedly likely to change the profile of this society. (...) We now have a proper grid enabling us to better understand what is going on in our societies in connection to disinformation and manipulation. (...) Looking only into the second level (manipulation, disinformation etc.) means failing to see the wood from the trees and turning the effects into causes and the other way around" (Dungaciu 2017: 12–13).

As such, I propose dismissing the established expression of "*information war*", in favour of "*semiotic war*". The first term conveys a naïve conception of communication, proper to common sense, in which the message is thought as prior and independent in relation to the process of communication, and the addressee is seen as a passive receptor, incapable of modifying the message in the interaction process (Fiske 1990: 1–4). This conception is allotted by Fiske to the "process school" from the field of communication theories. The process school is represented by Lasswell, Shannon and Weaver, Newcomb, Gerbner, Westley and MacLean, Jakobson (Fiske 1990: 24–35). It is also Fiske who argues that this concept is proper to the advertising companies' executives (Fiske 1990: 189). The second term proves the grasping of the fact that the message is the product of interaction and that the receptor is active, his/her activity decisively contributing to the production of the message. This approach is specific, according to Fiske, to the "semiotic school", represented by Ferdinand de Saussure, Charles S. Peirce, C. Ogden and I. Richards, L. Hjelmslev, Roland Barthes and Pierre Gouriaud (Fiske 1990: 46–60).

Reducing the semiotic war to a so-called "information war" is as if one would reduce an awareness campaign to an information campaign or the process of persuasion to a simple dissemination of information, as in the Shannon and Weaver model. There are, however, authors who, instead of the term "information war" proposed the term "information war"[15] which, although it fails to encompass everything, represents a step forward.

[15]See the contributions of scholars such as W. J. T. Mitchell, Bruno Latour and Peter Weibel at the online discussion of the concept of "image war", organized in *Nomadikon* by Øyvind Vågnes and Asbjørn Grønstad (http://www.nomadikon.net/contentitem.aspx?ci=320). In *Cloning Terror: The War of Images, 9–11 to the Present* W. J. T. Mitchell describes what he calls the image war thus: "it has been waged against images (thus acts of iconoclasm or image destruction have been critical to it); and it has been fought by means of images deployed to shock the enemy, images meant to appall and demoralize, images designed to replicate themselves endlessly and to infect the collective imaginary of global populations" (*apud* Mitchell 2012).

References

Addis, M., and M. Holbrook. 2001. On the conceptual link between mass customization and experiential consumption: An exposition of subjectivity. *Journal of Consumer Behavior* 1 (1): 50–66.

Andreopoulos, Andreas. 2006. *Art as theology: From the postmodern to the medieval*. London: Equinox Publishing.

Augustin, 2007. *On the true religion*. Bucharest: Humanitas.

Betea, Lavinia. 2000. *Communication and discourse in the "Wooden Language" of the communist regime*. Bucharest: Polirom.

Borţun, Dumitru. 2002. *Epistemic bases of communication*. Bucharest: Ars Docendi.

Borţun, Dumitru. 2011. Evolution of the public sphere in postmodernity: De-politization of political communication. *Communication et discours. Dimensions postmodernes—revista INTERSTUDIA*, nr. 10 (2), Editura Alma Mater Bacău, pp. 170–183.

Borţun, Dumitru. 2012. *Public relations and the new society*, 2nd ed. Tritonic: Bucharest.

Borţun, Dumitru, and Purcărea Victor Lorin. 2013. Marketing and semiotic approach on communication. Consequences on knowledge of target-audiences. *Journal of Medicine and Life* VI (1): 103.

Borţun, Dumitru. 2014. Mercerized communication between manipulation and ethics of responsibility. In *Authority and power of Christian values*, ed. C. Constantineanu, P. Runcan, and R. Runcan, 107–119. Bucharest: Didactic and Pedagogical Press.

Boudon, Raymond. 1997. Cunoaşterea. In *Tratat de sociologie*, ed. R. Boudon, 555–600. Bucureşti: Humanitas.

Ciapalo, Roman Theodore (ed.). 1997. *Postmodernism and Christian philosophy*. Mishawaka: American Maritain Association.

Chifu, Iulian, and Nantoi Oazu (eds.). 2016. *Information war: Standardization of informational aggression of the russian federation*. Bucharest: Publishing House of the „Ion I. C. Brătianu" Institute of Political Sciences and International Relations.

Chombart de Lauwe, Paul-Henry. 1970. *Images de la culture*. Paris: Petite Bibliotheque Payot.

Deely, John. 2001. *Four ages of understanding: The first postmodern survey of philosophy from ancient times to the turn of the twenty first century*. Toronto: University of Toronto Press.

Ducrot, Oswald, and Tzvetan Todorov. 1972. *Dictionnaire encyclopedique des sciences du langage*. Paris: Editions du Seuil.

Dungaciu, Dan. 2017. The triad of information war thinking. In *Information way under the microscope: Concepts, methodology, analysis*, ed. Lucian Dumitrescu. Bucharest: Editura Institutului de Ştiinţe Politice şi Relaţii Internaţionale „Ion I.C. Brătianu".

Eco, Umberto. 1976. *A theory of semiotics*. Bloomington: Indiana University Press.

Eriksen, J.-M., and F. Stjernfelt. 2010. Culturalism: From idea to unconscious presupposition. *Sociologija, II* 4: 359–376.

Fiske, John. 1990. *Introduction to communication sciences*, 2nd ed. London: Routledge.

Goian, Ion. 2017. Contemporary world between truth and post-truth. In *Encyclopedia of the fundamental works of political philosophy: Contemporaries: 2000–2017*. Bucharest: Publishing House of the „Ion I.C. Brătianu" Institute of Political Sciences and International Relations.

Iliescu, Adrian-Paul. 1989. *Philosophy of language and the language of philosophy*. Bucharest: Scientific and Encyclopedic Press.

Janik, Allan S., and Stephen S. Toulmin. 1973. *Wittgenstein's Vienna*. New York: Touchstone.

Kuhn, Thomas. 1970. *The structure of scientific revolutions*. Chicago: The University of Chicago Press.

Lee, M.S., H.D. Hsiao, and M. Yang. 2011. The study of the relationship among experiential marketing, service quality, customer satisfaction and customer loyalty. *International Journal of Organizational Innovation* 3 (2): 353–379.

Linton, Ralph. 1945. *The cultural background of personality*. New York: Appleton-Century Co.

Lipovetsky, Gills. 1996. *The Dawn of Duty. The painless ethics of the New Democratic Times*. Bucharest: Babel.
Marmo, Costantino. 2010. *La semiotica del XIII secolo [The semiotics of the 13th century]*. Milan: Bompiani.
Matei, Corina Sorana. 2016. *Postmodernity's fugitive truths*. Bucharest: Tritonic.
Merton, Robert. 1998. *Éléments de théorie et de méthode sociologique*. Paris: Armand Colin.
Mitchell, W. J. T. 2012. Image war. *Nomadikon*, 13, July 23, http://www.nomadikon.net/contentitem.aspx?ci=320.
Ogden, C., and Richards, I. 1923 [1949]. *The meaning of meaning*, London: Routledge and Kegan Paul.
Olteanu, Alin. 2009. *The semiotics of Saint Augustin, Bachelor's Degree*. Bucharest: Biblioteca Şcolii Naţionale de Studii Politice şi Administrative.
Olteanu, Alin. 2018. Semiotics as a proposal for a Numanistic educational programme. In *Readings in Numanities*, ed. O. Andreică and A. Olteanu, 3–17. New York: Springer.
Pârvulescu, Cristian. 2017. Low-cost Democracy in a Brave Illiberal World. *Sfera Politicii* 191–192: 3–14.
Peirce, C.S. 1931. *The collected papers of Charles Sanders Peirce*. Cambridge, Mass.: Harvard University Press.
Rorty, Richard. 1989. *Contingency, irony and solidarity*. Cambridge, New York: Cambridge University Press.
Revel, Jean-Francois. 2007. The need of ideology. In *The useless knowledge*. Bucharest: Humanitas.
Smilansky, S. 2009. *Experiential Marketing: A practical guide to interactive brand experiences*. London: Kogan Publishing House.
Smith, Adam. 1977. *An inquiry into the nature and causes of the wealth of nations*. Chicago: The University of Chicago Press.
Schmitt, B. 1999. *Experiential marketing: How to get customers to sense, feel, think, act, relate*. New York: The Free Press.
Thom, Francoise. 1987. *La Langue du Bois*. Paris: Julliard.
Wittgenstein, Ludwig. 1980. *The blue and brown books*. New York: Harper and Row.

Part II
Case Studies

Chapter 6
Towards a Holo-Semiotic Framework for the Evolution of Language

Jeremiah Cassar Scalia

Abstract The chapter develops the view that a global semiotic lens spanning biology and culture (thus covering the phenomenal processes of life generally) is a particularly vital frame for inquiry into the origin and evolution of language. It is further suggested that such inquiry, in its turn, could prove to be the single most important area of study to the work of transcending long-standing dichotomies between biological and cultural studies, and of affirming the perspectival salience of a semiotic theory of life. With respect to this view, a "holo-semiotic" framework (with roots in the global semiotic tradition of Thomas Sebeok) is developed wherein the object of the origin of language is approached and located within a semiotic coevolutionary complex of physio-anatomical force-relations (kinesio-/eco-semiotic) whereby a capacity for symbolical insight (anthroposemiotic) could phenomenally emerge out of biological impulse (bio-/zoo-semiotic). Based in the logic of a holo-semiotic model of language evolution, it will be argued that evolutionary semiotic processes hinge crucially on the inherent role that mimesis (as a principle function of semiosis and primary mode of information transmission) plays in both culture and biology—in its social imitative sense and in its biological replicative and adaptive senses.

6.1 Introduction

The origin and evolution of human language is widely considered one of the most difficult problems in the human sciences. Fundamentally, there are two reasons for this that stand out. Firstly, the emergence and subsequent evolution of the object of investigation took place deep in our paleolithic past and left nary a trace; there are after all no linguistic fossils to dig up, and thus no directly resultant evidence to analyze—a daunting challenge but one that is clearly recognized by all investigators. The second reason for difficulty pertains to the nature of inquiry and analysis

J. C. Scalia (✉)
Wenzhou-Kean University, Wenzhou, Zhejiang, China
e-mail: viacassar@gmail.com

A. Olteanu et al. (eds.), *Meanings & Co.*, Numanities - Arts and Humanities in Progress 6, https://doi.org/10.1007/978-3-319-91986-7_6

itself as it exists within academe, consisting specifically in the entrenched epistemological divisions that exist between the biological and social sciences—a division that language evolution study implicitly straddles. As such, contemporary language evolution scholarship lacks a clearly defined paradigm. As a body of literature it is a potluck of specialized theories, with scholars contributing from cognitive studies in linguistics and psychology, evolutionary biology, neuroscience, psychological anthropology, ethology, primatology, philosophy of language and mind, and even within these disciplines, approaches vary tremendously. These differing vantages on the problem have seen much disagreement and confusion not just over language's emergence but also over what is even meant precisely when we say "language". As biologist/cognitivist W. Tecumseh Fitch writes "all of us are still exploring the elephant of language in the darkness, all of us with only partial understanding" (2010: 2).

It is clear that going forward these divisions in academe must be adequately addressed and to the degree possible transcended if the problem is to be viably investigated, and if the significant challenges associated with the formerly stated reason for difficulty are to be met. It is in light of this fundamental requisite for an epistemological reconciliation that I turn to semiotics, the study of sign systems and processes, what John Deely has called "the only inherently interdisciplinary perspective there is" (cited by Kull 2014: 533). This is in keeping with Thomas Sebeok's assertion that the whole of nature must be understood under one mantle, as a "semiobiosphere", from the molecular code to the verbal code (Maran et al. 2012: 87). This conception has since been carried forward, and further developed and refined by biologist Jesper Hoffmeyer, among others, who has posited that:

> [...] the life sphere is permeated by sign processes (semiosis) and signification. Whatever an organism senses also means something to it – e.g., food, escape, sexual reproduction etc., and all organisms are born into a semiosphere, which is to say a world of meaning and communication: sounds, odours, movements, colours, electric fields, waves of any kind, chemical signals, touch etc. (cited by Hartley and Potts 2014: 140)

A global semiotic lens spanning biology and culture and thus covering the phenomenal processes of life generally, is certainly as vital to language evolution study as it is to any other interdisciplinary social/biological study, perhaps more so. Moreover, language evolution study in its turn may prove to be the single most important study to the work of transcending more broadly the aforementioned social/biological divisions in academe, and of affirming the perspectival imperative of a semiotic theory of life. Sebeok himself predicted that the "strategic anthroposemiotics/zoosemiotics dichotomy will stand, just as long as the riddle of the origin of human language remains unsolved" (Sebeok 1985: 299).

Conceiving of the world as a semiobiosphere, and thereby declaring the "anthroposemiotics/zoosemiotics dichotomy" to be false, is a rejection, if not explicitly stated, of a latent Cartesian mind/body dualism that many would argue still endures in academe. For it suggests that semiosis in its essence (the elemental set of aspects that define semiosic processes) is not bound to symbolic, representational communication as we understand it to be in the form of human language, thought and material culture, and that the symbolical currency of mind and the biophysical currency of body, distinct as they may be, are of a semiotic piece. It is from this holistic, global semiotic perspective that a framework for the origin and evolution of language is herein developed. This framework makes account of a number of distinct levels or fields of investigation: the biological/physiological (bio-/zoo-semiotic), the physio-mechanical (embodied, kinesio-semiotic), and the cognitive-symbolical (anthroposemiotic). This series of fields in holistic theoretic aggregate I refer to as *holo-semiotics* (from the Greek *holos* meaning whole) and is akin to Sebeok's "global semiotics" but would connote a more specific investigatory focus on Homo sapiens. Additionally, I will argue that evolutionary semiotic processes hinge crucially on the inherent role that mimesis [as a phenomenon of semiosis (Maran 2003), and even the primary mode of semiosis (Deacon 2004)] plays in both culture and biology, in its social imitative sense and in its biological replicative and adaptive senses. A faculty for mimesis will be further explored as a requisite preadaptation underlying a capacity for symbolic reference in language (Donald 1991).

This article will, in the following sections, first briefly consider the historical legacy of Cartesian dualism and its implications for language evolution research. Next a brief review of language evolution literature will be given. After this, the theoretical semiotic framework (holo-semiotics) incorporating a number of trans-disciplinary principles and perspectives will be laid out. This will be followed by a narrowing focus to several key concepts within this framework, which will lead finally to some concluding remarks.

6.2 The Cartesian Legacy and Its Implications for Evolutionary Linguistic Inquiry

> How debilitating is the view which dualises objects and mind, when the Buddha-body of Reality is free from individuated distinctions! How deluded we have been in clinging to the dichotomy between the environment and its inhabitants!
>
> —The Tibetan Book of the Dead

The word *semiosis* derives from the Greek root *sema*, which meant *sign*, but also *seed* which is likely the earlier antecedent. Typically, the notion of sign is associated with the human/symbolic world, but in a world understood as a semiobiosphere, signs (as in phenomenal signal processes) are something broader and deeper than human notions of representation, a sense that seems to resonate more powerfully with the earlier etymological meaning of seed. In its transcending—or transgressing depending on your point of view—of the divide between natural and human worlds and between biological and symbolical phenomena, a semiobiospheric perspective finds a certain congruence with a broadly familiar tenet that is inherent to a great many non-Western philosophical and religious traditions, cosmologies and practices which may be captured in the simple and somewhat reductive pronouncement: *mind and body are one*.

A germ of this basic tenet resides in the "Buddha-body of reality" of Tibetan Buddhism, as well as in the "yoking" of mind and body in the ancient Indian practice of yoga. It is present too in the "divine presence" of the Sufi mystics, in the swaying "ginga" of the Afro-Brazilian martial-art dance of capoeira, and further still in the experiential unity of the physical and spiritual realms of numerous animistic traditions throughout the world, all of these ascribing to some notion of a fundamental reality or spiritual home that is beyond the illusion of mind-body dualism. These traditional 'monistic roots' understandings of the human condition and experience (many of which are manifest as formalized embodied practices which are expressed through highly developed embodied articulatory inventories, e.g.—capoeira moves, yoga asanas) have long been characterized as 'primitive' or 'illogical' from the Cartesian perspective associated with Western culture and society and in which numerous contemporary dichotomous sets are rooted. Indeed, Western philosophy and science has been grappling with "the mind-body problem" since the mid-17th century when Rene Descartes reintroduced the age-old Platonic question and offered his own systematic theory of a metaphysical split between mind and body. Through the following centuries, the 'problem' would prove so compelling that it would come to define a fundamental and enduring philosophical episteme in explicit dualistic terms, extending well beyond meta-physical questions specific to mind and body and into all manner of emerging philosophical thinking, including the terms of analysis of newfangled specialized branches such as sociology, anthropology, psychology and linguistics as they emerged as fields in their own right through the latter half of the 19th century—with structural linguistics, as an essential kind of semiotic, becoming a defining if hidden analytical feature of all of them beginning in the early 20th century.

It wasn't until the latter half of the 20th century, that Western social science, particularly anthropology, began to faithfully resist and interrogate its own structuralist epistemic inheritance and open itself to the possibility of a wider, richer and more nuanced field of cultural and ethnohistorical knowledge, one that may include traditional indigenous and subaltern knowledge and experience not merely as worthy objects of examination as they had been since the early days of anthropology, but as indispensable tools of comprehension and understanding in themselves that must be brought into the enterprise of ethnographic description and the production of new social and anthropological knowledge (Clifford and Marcus 1986). The larger implications of this growing consciousness took hold in a crisis of academic representation, as a full-blown self-consciousness about the credibility of Western tools of analysis. There was some mysterious inescapable sense that something about Western positivist knowing was off; something was missing. This insecurity creeping across generations of scholars is what ultimately drove the reflexive turn, in which a critical mass of the new scholarship began to turn away from what had become by the early to mid 20th century a more structurally sophisticated but still relatively Cartesian form of dualism, e.g.—Levi-Strauss' structuralist "binary oppositions" thesis on culture which borrowed heavily from

Roman Jakobson's structural linguistics to understand universal structures of the "savage" human mind (Goody 1977: 4). Driven by post-modern and post-structural movements, this turn honed a more critical and reflexive analytic sensibility in the treatment of social and cultural phenomena. In more recent years this turn away from a neat binary positivism has been evident, at least in part, in varied neurobiological studies where data have appeared to affirm the turn away from the old binary and all but dualistic rigidity of social, cultural as well as behavioral and psychological analyses (e.g.—findings for complex cognitive–emotional behaviors having their basis in dynamic coalitions of networks of brain areas, none of which can be considered specifically affective or cognitive (Pessoa 2008).

So exactly what does all of this have to do with language? How pertinent is the question of language, or its evolution for that matter, to questions of human history and subjective consciousness, be they socio-cultural (including epistemological), or neurobiological? In the first place, what might be noted is the modern historical essentiality of structural linguistic analysis (with roots in Saussurean semiology) to all of the social sciences (including post-structuralist critique). Thus, of the social sciences, it was the language science that would guide and even dictate the terms of analysis of the rest, even extending its influence into physical and biological fields. Outside of the "purely semiotic system" of linguistics (Jakobson cited by Chandler, 2007/1929), this analytic approach to phenomena has reigned as a kind of shadow heuristic across the human sciences that, beneath the discursive surface of a given discipline's preferred lexicon, is implicitly semiotic. Couple this with the consideration that such consciousness shaped by the critical and specialized language of various fields of academia, as well as the concurrent evolution of popular consciousness in culture(s) at large, like every other cultural, ideological and epistemological consciousness before it, is naturally represented by and interactively carried out through the dynamic semiotic processing that is human language, the mimetic dia-logic of all psycho-socio-cultural phenomena.

If we are to inquire deeper into human language and its origins it may be necessary to more critically address not just how language may have become evolutionarily stable in paleolithic settings of the distant past but also how language has proceeded through an etymological history of the academy and served as the explicit edge of powerful dichotomous conceptions such as mind/body, nature/nurture, subject/object, social/psychological, etc., that have defined and distinguished a great myriad of fields and discourses, exponentially advancing but in some regards hamstringing our ability to approach certain subject matter. The great 20th century philosopher and linguist Ludwig Wittgenstein acknowledged something of this hamstringing when he suggested that the "mind-body problem" arises, quite simply, out of our linguistic capacity to describe human experience in different ways, i.e. with psychological, social or biological lexicon/vocabulary (1954). He reasoned that it is simply an illusory problem at the level of language where descriptions of human experience are based in different conceptual frameworks and the different signifiers (lexicon) used to describe them. The 'problem' then, does not really exist at the deeper signified level of reality. From such a vantage the phenomena of language can be understood to play a central and indeed causal role in the conceptual duality of mind and body among other binaries. Wittgenstein does not in this instance get into evolutionary causality though it should follow from this perspective that the origin of such a distinction is dependent upon the origin and evolution of language and symbolic thought, if indeed they are not ultimately one and the same origin. Thus approaching the question of language evolution may call for a critical inquiry into not only the paleoanthropological development of human language and culture in its natural communicative sense, but also the etymological development and semantic treatment of academic language, and the legacy of Cartesian thought. Because the devices that are used to investigate the evolution of linguistic phenomena are themselves linguistic, understanding how they emerged by gaining insight into their semantic seedbeds, may prove critical to a conceptual re-framing of the larger paleo-linguistic inquiry.

6.3 Language Evolution's Theoretic Axes: Sizing up the Elephant in the Darkness

Within the kaleidoscopic language evolution literature one can generally identify a core group of distinct and competing orbits of thought. Looming large among these are two sets of dueling theoretical oppositions. These two oppositional sets, seen biaxially, may form a coordinate system by which one might be able to place a given hypothesis with in its larger paradigmatic context; one axis may be referred to as the continuity/discontinuity axis, and the other as the neural/cultural axis.

Theories of evolutionary continuity suggest that language, given its complexity, must have evolved gradually in the normative Darwinian sense from earlier pre-linguistic/proto-linguistic systems with roots as far back as two million years ago. Discontinuity theory on the other end of this axis, for which the Chomskyan perspective has predominated, runs against this gradualist perspective, according to an argument that language is such a unique trait which cannot be compared to any other non-human species in the biosphere, and thus it must have appeared fairly suddenly during the course of human evolution.

As for the neural/cultural axis, neurally-based innatist theory, for which the Chomskyan view has also predominated, sees language mostly as an innately emerging faculty of the brain, and describes language's defining feature as a kind of computative language module made operative through a genetically encoded syntax. Some, like Pinker, might be described as continuity innatists, and others, like Chomsky, as discontinuity innatists. On the cultural end of this axis, theorists have pushed back against the Chomskyan view, regarding language as mainly an externally driven cultural system learned through social interaction over much longer swaths of evolutionary time. Taking innatist and discontinuity ideas together, the Chomskyan school of thought holds that language—as a language "organ" or module—emerged innately and suddenly through an abrupt change, a mutation in the brain of a single hominid which amounted to something like a faculty for language, likely occurring somewhere between 70,000 and 100,000 years ago (Bolhuis et al. 2014).

Weaving across these core axes is another important theoretical divergence which pertains to the question of language's original core modality of expression, with the predominating theoretical opposition being between modalities of speech/ vocal communication on the one hand and gestural communication on the other. Still further theories crisscross aspects of these foregoing theoretical polarities, and new and revised frameworks have emerged that find foregoing scholarship to be too often misguided by a singular, compartmentalized focus on language and a failure to recognize it as a subordinate aspect of the wider phenomenon of symbolic culture.

As multifarious as contemporary language evolution theory is, virtually all investigators across all disciplines do acknowledge to one degree or another the idea that the brain, and its neuro-cognitive processing, is central to the problem of language evolution. This has been explicit at least since the 1990s ("the decade of the brain" as designated by then U.S. president George H. W. Bush) which saw many important and exciting strides in brain research. Though, this neuro-cognitive/ neurolinguistic turn can be traced in earnest back to the 1960s when Noam Chomsky's Universal Grammar theory (UG) rocked the field of linguistics, and picked apart the behaviorist view which had long prevailed in the field. The enduring impact of Chomsky's theorizing of innate syntactic structures in the brain offering a universal neuro-architecture for language has been so profound that it can hardly be understated, even in an era in which prevailing and ascendant voices in the field have increasingly deviated from and/or disputed principles of Chomskyan thought. In his ideas we find the foundation for a human science paradigm that still endures, still defined as it is by a redoubled inward focus on neuro-cognitive computation. What is much murkier is just how such a "higher" cognitive computation is bound up with the autonomic nervous system, if or how cognitive computation can itself amount to a symbolical awareness/consciousness, and precisely what the theoretical terms are on which we may account for the evolutionary development of the object of study from a purely biological to a 'biological-plus-symbolical' constitution. Indeed, there has been and still is a profound theoretical challenge surrounding the question of symbolic reference, something that anthropologist Merlin Donald has likened to the "Achilles heel" of computational approaches to language. The difficulty is, in Donald's words, "to understand or use a symbol appropriately in context you must first understand what it represents, and this referential understanding is inherently non symbolic" (1991: 3).

A major challenge that remains is to more deeply probe and explicitly define the component features and dynamics of the "inherently non-symbolic" in non-negative terms. This, I will argue, brings us unavoidably to consider the only domain of meaning that can be said to undergird and intertwine with symbolic meaning: the embodied, affective, physiological domain. Investigating the terms of meaning at this level is not altogether resistant to a computational approach. It is rather more a question of situating computational modeling within a global semiotic hermeneutical frame as opposed to an exclusively cognitive linguistic one.

6.3.1 Coevolutionary Perspectives

Among the theories that have countered the Chomskyan view of language evolution, a significant theoretical strain that bears mention consists in an array of perspectives grounded in coevolution. It is certainly true that many theorists simply do not consider coevolution to be a crucial concept to understand or approach their object. Still, a growing number of theorists intuiting the conceptual salience and generative primacy of coevolution have taken up with a kind of coevolutionary ontology for language. Notable coevolutionary theories include language and brain coevolution (Terrence Deacon, Steven Pinker), speech and ritual coevolution (Roy Rappaport, Luc Steels, Terrence Deacon) and speech anatomy and brain coevolution (Philip Lieberman).

In The Symbolic Species, Terrence Deacon endeavors to uncover "the logic linking cognitive functions to brain organization" and develops a theory of an emergent metaphysical "social contract" that is integral to the coevolution of language and brain (1998: 25, 408). Philip Lieberman, who stands in particularly rigid opposition to the innatist modular view of the Chomskyan camp as mentioned above, proposes a coevolutionary process beginning some two million years ago, when Homo first began to branch off from its closest primate ancestor. Though, instead of focusing on the brain's coevolution with "language", symbolic thought or culture, Lieberman is much more focused on speech and its physio-anatomical hardware, believing as he does that the evolution of speech is the key to the evolution of language in general (2002: 36). Similar to these, the global semiotic framework to be elaborated in the next section assumes an essential coevolutionary ontology, and subscribes to the notion that, if language is a complex adaptive system, as I believe it to be, then a given hypothesis for language evolution must reckon with coevolution, since all evolution is coevolution (von Wahlert cited by Schulze 2012: 450), the ontological essence of which is echoed in a second truism to which I would also subscribe—that all adaptation is exaptation (Dennett 1996: 281).

6.3.2 Reexamining the Evolution of Syntax in the Context of Motor Control

I would add a final word about syntax and the question of its core significance to language and its evolution. According to the innatist Chomskyan view, syntax is what makes language so special (species-specific to Homo sapiens). And the defining characteristic of syntax—which is said to separate it from other ethological communication systems of the animal kingdom, and even perhaps from other functional cognitive systems in the human brain—is what's known as recursion, which is the embedding of a function (sentence structure) within another function (sentence structure) of the same type. This is fundamental to Chomsky's notion of Universal Grammar (and later, Merge) allowing human language to make infinite use of finite means. Though as Lieberman has pointed out, there is evidence that this is not in fact unique to linguistic function. Motor actions such as walking, running, climbing or dancing operate through selectional constraints which "yield hierarchical patterns… that potentially can generate an infinite number of combinations" (2015). This is consistent with the longstanding assertion by Karl Lashly that neural mechanisms originally adapted for motor control constitute the bases of syntax and human creative behavior (Lieberman 2007: 39). More recent findings in neuroscience, Lieberman notes, reveal a close intertwining of syntax, cognition and speech motor control, all of which have shown to be in deficit in a specific speech disorder traced to the FOXP2 gene, a protein that is required for the proper development of speech and language (2007: 52). This close link between syntax and motor control function, which Lieberman has shown to be capable of a kind of corporeal recursion hierarchically similar to a syntactic linguistic recursion, suggests that there is a possibility that abstract mental grammar may be built off of, or mapped from, a motor-activated corporeal infrastructure, i.e.—the body.

6.4 Towards a *Holo-Semiotics* of Human Language

The holo-semiotic perspective herein developed should comport well with evolutionary linguistic theory that views human language as a complex adaptive system operating on multiple interacting time scales accounting for biological evolution and cultural transmission (Kirby 2017: 118, Wang 2010: 2). As such language is understood to be a product of Darwinian evolution and natural selection (though with critical uniquenesses explored further along), and, like all other adaptations in nature, finds its evolutionary and therefore existential merit in its degree of correspondence to the demands of survival. It is further maintained that, as powerful and unique as it is, the cognitive linguistic competence emergent from the gelatinous conglomeration of reticulated neural folds that constitutes the human brain is nonetheless bound to an order of 'semiotic evolutionary competence' in which it shares company with that part of the honeybee's autonomic nervous system that can communicatively map directions to pollen sources by doing a butt-waggle dance—something that has been shown to be a simple but semantically compositional system communicated in analogue fashion (Hurford 2008: 7), with the dung beetle's navigation capabilities that operate on polarized moonlight (Dacke et al. 2003), with the multitudinous examples of brilliant reticulated coloration patterns that mark the outer skins and coats of mammals, reptiles and insects alike. All of these serve behaviorally and/or anatomically as an inter-semiotic complement to a species' biological niche.

An understanding that finds such zoo-semiotic features (and feats) to be on the same continuum, and thus within the same investigatory paradigm, with cognitive symbolic capacities of humans (anthroposemiotic) extends from a conception outlined by Jesper Hoffmeyer and Frederik Stjernfelt which posits that if life and semiosis are fundamentally co-extensive, an attempt may be undertaken to "classify cognitive and communicative potentials of species according to the plasticity and articulatory sophistication they exhibit" (2016: 7). It is argued that a classificatory approach to communicative potentials which takes as an organizing principle the "plasticity and articulatory sophistication" exhibited by species, is an especially cogent way to approach human language as a complex adaptive system. With this as a defining principle, human language and its evolution is approached from what may be referred to as a *holo-semiotic* perspective consisting in biosemiotic, zoosemiotic and anthroposemiotic domains.

6.4.1 Locating the Object of the Origin of Language

Proceeding from a global "semiobiosphere" vantage, or what I am now attempting to develop as a holo-semiotics, the essential driving question to be pursued is: how and why did human language evolve the way that it did? That is, how can we account for the phenomenal emergence of an abstract, human language-based social consciousness out of the unreflective autonomic consciousness associated with biological instinct? To pursue this inquiry I approach the object of the origin of language locating it within a coevolutionary complex whereby:

(1) **biological impulse** (bio-/zoo-semiotic), operating through certain
(2) **physio-mechanical force-relations** (kinesio-semiotic), deployed against the
(3) **environmental constraints** that ever-define an organism's peri-personal area (ecosemiotic) could give rise to
(4) **symbolical insight** (cultural/anthroposemiotic).

This framework facilitates a unified perspective on a coevolutionary process that may proceed through qualitatively different semiotic levels or zones. This approach generally coheres with contemporary frameworks that have taken root in biosemiotics following in the global semiotic tradition of Sebeok, as represented in the work of Kalevi Kull and Jesper Hoffmeyer. It further shares at least a periphrastic relation with the classical semiotic categories of Firstness, Secondness and Thirdness put forth by Charles S. Peirce (Potter 1967: 16). Kull's conception of "semiotic threshold zones" (2009: 8) are particularly instructive for the development of a holo-semiotic framework for language evolution. It further calls on us to reexamine the body, with its own articulatory dia-logic by which it interfaces with and navigates all that it is outside of itself, i.e.—Jakob Von Uexküll's inter-semiotic cybernetic concept of *innenwelt* (an organism's inner physiological self) and *umwelt* (the surrounding environs as interfaced by the organism), and which may "denote the subjective (subjectivized, meaningful) world of an organism" (Kull 2001: 7).

Kull's threshold zones, which define three general semiotic zones by which qualitatively distinctive life processes may be classified, include: the vegetative threshold zone (iconic), the animal threshold zone (indexical), and the cultural threshold zone (symbolic) (2009: 15). Within the framework that I've outlined above, (1) *biological impulse* and (4) *symbolical insight* would correspond respectively with Kull's animal threshold and cultural threshold and with classical Peircean notions of Firstness (feeling/potential) and Thirdness (representation). The space where (2) *physio-mechanical force-relations* meet the constraints of the (3) *peri-personal environmental zone* that an organism at all times finds itself within, is understood as a synchronic zone of (inter)action, and generally corresponds with Secondness (action/reaction). Together, points (2) and (3) make up the real time micro-dynamics of living, adaptation's ever-present ground zero which fundamentally directs the diachronic trajectory of macroevolutionary processes. This ground zero field I refer to as the bio/eco interface which itself can be divided into three general zones, each one more concentrically remote (from the organism's subjective center) than the previous, and is thus defined by a core/periphery structure:

(a) **bio-sensory brain** (core)
(b) **bio-sensory superficies of the body/nervous-system** (peripheral)
(c) **physical environment** (extra-peripheral).

This core/periphery network structure is important to understand if you take to heart research that indicates that "the topological structure of networks is necessary for understanding their internal organization, particularly if they are continuously evolving and changing over time" (Otokura et al. 2016: 204). Thus, if a biologically stable system's core and periphery may be understood to be mutually constitutive, we might, from a coevolutionary perspective, inquire into the complex workings of that system (i.e.—the brain-nervous system continuum that is the body) as it interfaces with its extra-peripheral surrounding environs. This may lead us to consider the degree to which critical changes in the brain may be coevolutionarily bound to the sensorial and kinesio-semiotic competencies of distinctive parts of the body's anatomical periphery.

It is a basic fact of biology that energy is stored by cells, embedded in the molecular structure of carbohydrates, lipids and proteins which will release that energy when introduced to oxygen through the process of respiration. It is respiration that allows the body to access and wield the energy that it has stored. Considering the circulatory system's relay of energy (blood circulation) in the context of the body's peripheral network of appendages, which are essentially the brain's biosemiotic intermediaries (head, face, ears, eyes, limbs, larynx, lips, tongue etc.) with the surrounding environs, we might then ask how this network's sphere of evolutionary influence could have been structured in both hierarchical and symbiotic senses (i.e. modeling the body's symbiotically-grounded hierarchy of semiotic affordances); that is, how did the various parts of the network interrelate (co-evolve) and did the affordances of some parts of the network come to exert a more significant force on the ultimate trajectory of the evolutionary process than others? Could clues to such a question be evidenced in the very structure of the nervous system and its circulatory pathways? This is a critical line of inquiry into how the "semiotic work" of survival during a critical evolutionary period might have been delegated to various parts of (b) the nervous system's peripheral network, that which constitutes adaptation's intermediating liminal zone between (a) the core and (c) the extra-periphery, which has implications for the evolutionary pathway from (1) biological impulse to (4) symbolical insight.

6.5 Mimesis by Analogue: An Adaptational Path Forward

There is something in the very foundation of our nature that puts a premium on the capacity to copy as well as to code, from DNA replication to natural selection (Barbieri 2008: 577). These two interrelated mechanisms, copying and coding, crucial genetically and evolutionarily, call on us to reexamine mimesis vis a vis semiosis, in biological (biosemiotic) and cultural (anthroposemiotic) contexts, and certainly in the context of a developmental bridge taking us from the former to the latter.

Timo Maran points to the problem of mimesis as conceptually critical to "the formation of new structures by semiosis as well as the development and changeability of semiotic systems" (2003: 211). Maran bases "the creation of mimesis" in a "recognition of the mimetic potential of the perceived object, situation, event, phenomenon or person by the creative subject." This recognition is the prerequisite to "the activity of expressing, revealing or performing this potential mimetically" (2003: 199). Maran's semiotic negotiation of mimesis[1] is undertaken at a cultural (anthroposemiotic) level of human imitation. If mimesis, in as far as it implies iconicity, is indeed an inherent function of semiosis, or even the primary mode of information transmission (i.e. semiosis) (Deacon 2004), it would invite a considerably deeper exploration of mimetic processes in the realm of bio-semiosis, in which a particular focus on evolutionary processes may prove critical to the question of language origins.

What is called "recognition of mimetic potential" in cultural terms may be understood in Darwinian evolutionary terms to parallel a selection for mimetic potential when mimetic correspondence favors survival. Thus the "perceived object" of a cultural mimesis may simply be referred to as an environmental *source-object* in the context of bio-evolutionary mimesis. The "creative subject" then, is no subject at all but the naturally selective evolutionary process that over time and

[1]Maran goes on to reference a potential parallelism between notions of semiosis and mimesis, citing Charles Morris who defined semiosis as a sign process consisting in "that which acts as a sign, that which the sign refers to, and that effect on some interpreter" (2003: 211). This parallelism is assumed but not extensively explored in this article, and going forward should prove critical to the development of semiotic theory of language and its origins.

across innumerable generations of a species may culminate in a kind of mimetic physio-anatomical and/or behavioral form. The question then is precisely what does such a mimetic selection turn on; what does the process operate through?

To approach this, I have turned to the concept of analogue, arguing that without a certain critical level of analogical correspondence (i.e.—critical mass of signal parallels), between an agent and its object, mimesis cannot successfully occur—whether in nature or culture. In nature, a *mimesis by analogue* would be associated with a natural selection process whereby successful mutations trend toward a high mimetic correspondence to a particular environmental source-object (or series of co-auxiliary source-objects), adaptationally culminating in an analogous form that generally mirrors the appearance of the source-object in structure, texture, color and/or behavior, e.g.—a stick bug that comes to look like a stick and thus is better shielded from predation; a polar bear's white coat that, via a molecular coloration analogue, mirrors its surrounding white tundra, thus better shielding its predatory approach to its prey. In this way a mutation may be selected for based on a mimetic correspondence with a source-object that may be in someway critical to an organism's survival; such mimetic phenomena is fundamentally analogical, its potential being defined by the presence of a critical confluence of physio-anatomical analogues. Just as a polar bear's white fur mutation would become adaptationally salient as a molecular coloration analogue of the white tundra, thus leveraging a selection for white fur polar bears, we might also consider: if language is an adaptive system, and if it is potentially the culmination of a multi-layered adaptive mimesis fundamentally shaped at each stage by analogical dimensions, what is the primary (and primal) analogue that it could have been anchored by, the essential physio-biosemiotic analogue through which a new sphere, the sphere of anthroposemiosis, could have emerged?

More concretely, the essential question being asked is: what, if there could be one, is the environmental source object(s) for language? It is argued that this question is central to the investigation of language's evolutionary biological emergence. To date this question remains unanswered, and as far as one can tell has not even been asked in these explicit terms. One might question the wisdom of searching the space of the material/physical world for an adaptational source object to which an immaterial/metaphysical complex adaptive system could be bound. And it is certainly a fair question. A dialogue should be had over it. What I would suggest in the first instance of such a dialogue is that the hermeneutic footing from which this kind of questioning springs is demonstrative of the continuing (albeit latent) influence of a fundamentally Cartesian orientation within Western academe, where the material (body) and immaterial (mind) dimensions of human experience are still largely irreconcilable, notwithstanding the great efforts to shrink the divide, by more innovative trans-disciplinary empirical inquiry and a more nuanced discursive finesse. As long as inquiry into language origins remains unduly constrained by disciplinary convention—which is itself secured by the compartmentalist language that expresses the convention, we shall not find the imagination nor the operative lines of inquiry to truly break new ground, and to adequately deal with the possibility that material-physical and immaterial-symbolical phenomena may be bound up together in a discoverable way; indeed we shall continue to be constrained, consciously or not, by the "strategic anthroposemiotics/zoosemiotics dichotomy" that Sebeok so long endeavored to transcend.

Further developing conceptual frameworks for mimesis and its processual terms as it occurs at zoosemiotic and anthroposemiotic levels is critical to any approach to the "riddle of the origin of language". It is hoped that the notion of something like "mimesis by analogue" may open up new avenues of inquiry for language evolution vis a vis zoological evolution, prompting new questions that may seek out core principles that pervade all areas and levels of biological life (including human) and thus make the terms of a phylogenetic trajectory across them traceable.

6.6 Conclusion

The contemporary theoretic axes that define investigations into the origins of human language, culture and cognition contain a very diverse array of approaches. Within much of the discourse, application of semiotic theory remains limited (with some exception) functioning more as a rhetorical auxiliary or stopgap perspective than as a critical, indispensable lens through which to view the subject matter and its related phenomena. In this paper I have presented a case for the latter application, arguing that the conceptual bases of semiotics afford a much needed unitary framework from which to explore the problem of language origins—a problem which straddles biological and symbolical (as well as ethological and cultural) domains of investigation.

The "holo-semiotic" framework that's been presented makes account of a number of distinct levels or fields of investigation: the biological/physiological (bio-/zoo-semiotic), the physio-anatomical (embodied, kinesio-semiotic), and the cognitive-symbolical (anthroposemiotic). From within this framework, it's been argued that the evolution of language must have been defined by a mimetic dimension, and that an "adaptive mimesis" antedating language's emergence would have been dependent on the presence of analogical correspondences at a physio-anatomical level. That is to say, the mimetic coevolutionary symbiosis ultimately culminating in human language must have derived from a kind of pre-cultural, pre-symbolic analogical relationship between covariant aspects of wholly biophysical phenomena becoming increasingly correlated through the process of natural selection.

There are indeed a number of contemporary perspectives in the language evolution scholarship as discussed above that cut trans-disciplinarily across the biological and social sciences, and that have, quite encouragingly, revealed new and intriguing points of coherence, and the depth of this coherence may yet show to be even more profound. In effort to navigate this deeper coherence, the notion of "holo-semiotics" developed in this article initiates a new synthesis of relevant disciplinary perspectives, principles and evidences that, it is hoped, might bring more clarity of vision to a perennially difficult problem.

References

Barbieri, Marcello. 2008. Biosemiotics: A new understanding of life. *Naturwissenschaften* 95 (7): 577–599.

Bolhuis, Johan J., Ian Tattersall, Noam Chomsky, and Robert C. Berwick. 2014. How could language have evolved? *PLoS Biology* 12 (8): e1001934.

Chandler, Daniel. 2007. *Semiotics: The Basics*. Routledge.

Clifford, James, and George E. Marcus. 1986. *Writing Culture: The Poetics and Politics of Ethnography*. University of California Press.

Dacke, M., D.E. Nilsson, C.H. Scholtz, M. Byrne, and E.J. Warrant. 2003. Animal behaviour: Insect orientation to polarized moonlight. *Nature* 424 (6944): 33.

Deacon, Terrence. 2004. Memes as signs in the dynamic logic of semiosis: Beyond molecular science and computation theory. *Conceptual Structures at Work*, 236–236.

Deacon, Terrence. 1998. *The symbolic species: The co-evolution of language and the Human Brain*. New York: W. W. Norton & Company.

Dennett, Daniel C. 1996. *Darwin's dangerous idea*. New York: Simon & Schuster.

Donald, Merlin. 1991. *Origins of the modern mind: Three stages in the evolution of culture and cognition*. Cambridge, MA: Harvard University Press.

Fitch, W. Tecumseh. 2010. *The evolution of language*. Cambridge, UK: Cambridge University Press.

Goody, Jack. 1977. *The Domestication of the Savage Mind*. Cambridge University Press.

Hartley, John, and Jason Potts. 2014. *Cultural science: A natural history of stories, Demes, knowledge and innovation*. London, UK: Bloomsbury Publishing.

Hoffmeyer, Jesper, and Frederik Stjernfelt. 2016. The great chain of semiosis. Investigating the steps in the evolution of semiotic competence. *Biosemiotics* 9 (1): 7–29.

James, Hurford. 2008. *Animal syntax? Implications for language as behavior*. Scotland: University of Edinburgh (ms.).

Kirby, Simon. 2017. Culture and biology in the origins of linguistic structure. *Psychonomic Bulletin & Review* 24 (1): 118–137.

Kull, Kalevi. 2009. Vegetative, animal, and cultural semiosis: The semiotic threshold zones. *Cognitive Semiotics* 4 (Supplement): 8–27.

Kull, Kalevi, and Ekaterina Velmezova. 2014. What is the main challenge for contemporary semiotics? *Σημειωτκή-Sign Systems Studies* 4: 530–548.

Kull, Kalevi. 2001. Jakob Von Uexkull: An introduction. *Semiotica-La Haye Then Berlin-* 134 (1/4): 1–60.

Lieberman, Philip. 2002. On the nature and evolution of the neural bases of Human language. *Yearbook of Physical Anthropology* 45: 36–62.

Lieberman, Philip. 2007. The evolution of Human speech: Its anatomical and neural bases. *Current Anthropology* 48 (1): 39–66.

Lieberman, Philip. 2015. Language did not spring forth 100,000 years Ago. *PLoS Biol* 13 (2): e1002064.

Maran, Timo, Dario Martinelli, and Aleksei Turovski. ed. 2012. Readings in zoosemiotics. *Semiotics, communication and cognition*, vol. 8. Berlin: Walter de Gruyter.

Maran, Timo. 2003. Mimesis as a phenomenon of semiotic communication. *Σημειωτκή-Sign Systems Studies* 1: 191–215.

Otokura, Mari, Kenji Leibnitz, Tetsuya Shimokawa, and Masayuki Murata. 2016. Evolutionary core-periphery structure and its application to network function virtualization. *Nonlinear Theory and Its Applications, IEICE* 7 (2): 202–216.

Pessoa, Luiz. 2008. On the relationship between emotion and cognition. *Nature Reviews Neuroscience* 9 (2): 148.

Potter, Vincent G. 1967. *Charles S. Peirce on norms & ideals*. Amherst, MA: University of Massachusetts Press.

Sebeok, Thomas A. 1985. Zoosemiotic components of communication. In *Semiotics: An introductory anthology*, ed. Robert E. Innis, 294–318. Indiana, USA: Indiana University Press.

Schulze, Ernst-Detlef. 2012. *Flux control in biological systems: From enzymes to populations and ecosystems*. London: Academic Press.

Wang, William S-Y. 2010. *Conference on Evolutionary Linguistics II*, Nankai University. Incomplete preliminary draft presented May 2010.

Wittgenstein, Ludwig. 1954. *Philosophical investigations*. New York: Macmillan.

Chapter 7
Multimodal Propositions and Metaphors in the Movie *Submarine*: An Application of Peirce's Doctrine of Dicisigns

Alin Olteanu

Abstract This chapter integrates Charles Peirce's doctrine of Dicisigns in the cognitive semiotics of film. It makes use of Peirce's concept of proposition to develop a framework for the analysis of multimodal meaning construction. As such, this theoretical addition to the cognitive semiotic theory of film illustrates how information is conveyed and, also, how arguments and metaphors are developed multimodally in the cinematic medium. The theory is further on developed and illustrated by an analysis of the movie *Submarine* (Ayoade in Submarine, 157, 2010). Particularly, the chapter discusses how propositional structures are formed through cinematic techniques and how the two metaphors "light/darkness is life/death" and "submarine biology is submarine psychology" participate in the unfolding of the movie's narrative.

7.1 Introduction

The present chapter has the twofold purpose of (1) contributing to the development of a cognitive semiotic framework for multimodal analysis by (2) offering an analysis of fragments of the movie *Submarine* (Ayoade 2010). The article does not develop a film theory of its own and it does not claim to exhaustively analyse the movie. Rather, it is a contribution to the cognitive semiotics of film, illustrated by analysing aspects of a film.

Warren Buckland makes the case for and developed a cognitive semiotic approach to cinema (2000). He argues that cognitive semiotics overcomes rationalist assumptions that survived in other contemporary film analysis methods, including the classical semiotic approaches of structuralist inheritance and cognitivism, which present "problems of translinguistics, the behaviorism of modern film theory's account of subject positioning, together with the cognitivists' overemphasis on the spectator as an autonomous rational self" (2000: 13). Thus, the

A. Olteanu (✉)
Kaunas University of Technology (KTU), Kaunas, Lithuania
e-mail: Alin.olteanu@ktu.lt

A. Olteanu et al. (eds.), *Meanings & Co.*, Numanities - Arts and Humanities in Progress 6, https://doi.org/10.1007/978-3-319-91986-7_7

epistemological virtue of semiotics is that it offers a frame for multi- and cross-modal transduction, untainted by strictly linguistic translation theories.

The present contribution to Buckland's cognitive semiotic film analysis method is made by using Stjernfelt's account of Charles Peirce's doctrine of Dicisigns (2014), exemplified by analysing the movie *Submarine*. While bringing more recent developments of cognitive semiotics to Buckland's theory, I also challenge his positioning of Peirce in the "Language Analysis tradition", together with the analytic philosophers, Saussurean structuralists and Habermasian pragmatists (2000: 15). Buckland's reading of Peirce as aiming "to transform questions about epistemology and the mind into questions about language and meaning" (2000: 15) also results in ignoring Peirce's anti-psychologism, making his semiotics a development of Cartesian introspection and Kantian epistemology (2000: 16). Stjernfelt argues that unlike other cognitive accounts of multimodal representation, such as those of Barsalou (1999) or Eco (1997), Peirce's doctrine of Dicisigns does not endorse a psychologistic explanation for multimodal perception (Stjernfelt 2014: 35–36).

It is at the crux of Peirce's pragmatism that meaning is accessible by our translation from "a confused form of thought" into "a sentence in the indicative mood" (CP 5.18). This is relevant because a Dicisign (proposition) is "equivalent to a sentence in the indicative mood" (CP 2.315). Thus, regarding cinema, Peirce's doctrine of Dicisigns accounts for our primarily propositional interpretation of filmic events. Upon this propositional layer of signification, which, as I explain below, frames the deictic dimension of a film, further interpretation is made available.

In this vein, Peirce's semiotics proves even more favourable to contribute to a cognitive semiotic film analysis, such as Buckland's. Brandt (2005: 1581) suggests that psychologistic theories of meaning cannot avoid the theory of language as double articulation and, consequently, in their approach to logic cannot grasp the meaningfulness of possible worlds. Double articulation, or double partterning, is the hypothesis in linguistics, starting mainly with de Saussure (1959 [1915]), that meaning is a linguistic double articulation of form and content. The term was first used by Martinet (1962) in his study of language as the phonetic articulation of sounds. In various forms the hypothesis became established as the norm in linguistics and some accounts of social and cultural semiotics (e.g. Hjelmslev 1954; Jakobson 1990; Chandler 2002: 6–7). This view is challenged by Peirce's semiotics which rather endorses a pluralistic, in contrast with dualist, construal of meaning. Also, the double articulation hypothesis was recently challenged in social semiotic scope by Kress and van Leeuwen (2001: 4).

As such, psychologism, given its analytic inheritance, faces a serious difficulty in accepting the contextual truth of fictional discourses: fiction can only be understood as *lying*. For instance, Brandt makes the case that mental space theory, by bringing claims for logic starting from cognitive representations, treats meanings "either as truth or as mental events" (2005: 1581). As such, it construes fiction as a purely mental event (a "headache", as Brandt describes it), outside the domain of *truth*. Peirce's semiotics is not problematic in this regard, particularly given his concepts of (1) real possibility (see CP 5.547), holding that possibility is a mode of reality and of (2) Universe of Discourse (see CP 6.531), claiming that truth is attributed to contextual discourses which are continuous in a (pan-)semiotic reality. Accordingly, Stjernfelt (2014) argues that Peirce's notion of dicisign affords truth value to statements within fictional discourses. Brandt comprehensively explains these advantages of the (particularly) cognitive semiotic approach to fiction:

> The main difference between a cognitivistic and a truly cognitive approach is thus that the former only intends to logically justify the semantic paradoxes, metaphors, counterfactual conditionals, and so on—by showing that they are still in principle possible, and virtually meaningful—whereas the latter intends to obtain a full-scale study of their reality, including discussions of the structural and functional grounding of cognitive and affective productions in the architecture of the socially and culturally committed human mind and of its pragmatic and semiotic dispositions. Analytic cognitivism—the historical origin of [Mental Space Theory]—is essentially but a logico-philosophical representationalism. By contrast, cognitive semantics, and especially cognitive semiotics, intends to study how representational meaning can be modeled as dynamically related both to the imaginary and to the experiential world in which meaning is meaningful (Brandt 2005: 1585).

Particularly the movie *Submarine* is chosen for this purpose for a number of reasons. Being based on a book (Dunthorne 2008), it illustrates psychological states through a narrative by audiovisual, kinetic means. The inner states of the characters (feelings, emotions, thoughts) are analysable in an anti-psychologistic framework, which observes strictly what the movie conveys as a discourse, avoiding any psychological speculation. By this it is posited that the beholder of the movie does not need to understand what were the director's thoughts in order to understand what the fictional characters are claimed to experience. This semiotic perspective is in tune with Kress' semiotic studies on multimodality (Kress and van Leeuwen 2001; Bezemer and Kress 2008; Kress 2010) and it can be used as a contribution to such hermeneutic. Moreover, by its emphasis on psychological states, the movie has a significant poetical dimension: not only that psychological states are conveyed via metaphors but the movie illustrates the characters' self-creation. Brandt and Brandt (2005) argue that "cognitive poetics should encompass not only poetry but also prose, drama and any hybrid genres, or 'modes' of writing" (124) because "it seems sensible to include in our notion of 'poetics' all literary forms of writing, in the spirit of the etymological root of the word (thus, the Greek word poiesis, creation, refers to all creative uses of language). Adopting this latter view, the term 'cognitive poetics' refers to cognitively-oriented generalizations on creative (read: literary) writing as such and can be used interchangeably with 'cognitive literary studies' to indicate the study of literary creations in a cognitive perspective" (124). From this perspective, a cognitive semiotic approach reveals the poetic dimension of *Submarine*, given the meaning dynamics by which some characters create their own personality.

I argue that cognitive semiotics, grounded in Peircean semiotics, offers a holistic hermeneutics for multimodal media. By holistic hermeneutics is meant a philosophical framework that is not subject to modern dualistic dichotomies of media and modalities, such as text/image, and of the sciences, such as logic/art. The key elements of the analysis stand in Charles Peirce's multimodal account of the sign.

Besides the recent developments in multimodal semiotics, I also draw on the biosemiotic concept of the body (Stjernfelt in Nöth 2006; Hoffmeyer 2008), and the cognitive semiotic, non-dualistic understanding of language and experience (Brandt 2005; Brandt and Brandt 2005). By explaining via these concepts how the cinematic techniques employed in this movie construct the narrative alongside with characters' subjective experiences, the semiotic framework proves a fertile hermeneutic for analysing art (and not only) in a postmodern, non-dualistic key. The psychological processes which the movie's characters undergo are discussed without psychological speculation, but only by considering the construction of the discourse by multimodal means. As such, this inquiry contributes to developing a philosophical understanding of contemporary art on its own multimodal terms. While everyday life and its artistic expressions, arguably, moved on to a postmodern stage, where mind/body dualism and substantialist ontology are not of concern anymore, philosophy still did not develop a hermeneutic paradigm fully detached from modernism. Simultaneously, an ongoing change of media reshapes society. Kress and van Leeuwen (2001: 1) observe that various media tend to move away from monomodalism. This suggests that multimodal expression is a feature of non-dualism. The hermeneutics stemming from modern dualist philosophy, itself monomodaly expressed (in plain text), is not designed to approach multimodal media. Given its dismissal of ontological dualism, semiotics supports such a framework for analysis.

Peirce's semiotics has been was already proposed by some scholars as appropriate postmodern philosophy (Deely 1982, 2001, 2009). Nevertheless, a non-Peircean branch of semiotics has been rather employed for cultural and art analysis. Arguably, this branch of semiotics, of structuralist and poststructuralist origins, still inherits a perspective on language and meaning as "double articulation" (de Saussure 1959 [1915]: 112–113, Chandler: 6), which underpins a dualist stance. As such, by detaching form from content, this perspective encounters some difficulties in approaching multimodal expressions and cross-modal translation. This is clearly observed, for instance, in de Saussure's idea, stemming from his divorcing of modalities (i.e. speech and writing) into altogether different sign systems, that more can be learned about someone by direct observation than by beholding a photograph (1959 [1915]: 23–24). Certainly, by having someone present, in the spatiotemporal sense, we can inquire from her about herself, but I consider that real existence, direct observation is more elusive than a photograph or a video. As I shall explain, contrary to de Saussure's argument, in *Submarine* stills are often used to convey the inner states of characters. The need for cross-modal translation, termed also *transduction* (see Bezemer and Kress 2008), raisesthe awareness that meaning is implicit in form. Morphology and syntax construe and, to various extents, contain meaning.

On Peirce's account, the sign-relation is not bound either to mental or non-mental realms but, rather, using Deely's term, it is suprasubjective (see Deely 2001). As it occurs in several of Peirce's definitions of the sign, each of the sign's termini, can be *anything*:

> Now a sign is something, A, which denotes some fact or object, B, to some interpretant thought, C. (CP 1.346)

On this account of signification as suprasubjective, the form/content dichotomy is also avoided. The Peircean framework, as considered a proper postmodern philosophy because of its non-dualism, proves a fitting approach to contemporary artistic productions. I mean contemporary in a loose temporal sense, as recently and currently produced, belonging to the digital age, and not in an art history sense, as referring to a certain artistic school. Also, the analysed movie is not labelled as 'postmodern art'. Nevertheless, as Kress and van Leeuwen remark, artistic expression is not aiming at monomodality anymore. Cinema is by default a multimodal medium and it therefore requires an adequate philosophical comprehension. The movie *Submarine* serves as a good example for illustrating the hermeneutical resources of semiotics, which, in turn, can bring to light non-modern aspects of artworks. The main reason for this is that this movic (likc many othcr artworks might) is particularly aimed at depicting the inner subjective experiences of characters. It does so through embodied actions and multimodal articulations consisting of a bringing together of dialogue and monologue with sceneries, frames, colors, visual contrasts and music. Thus, the movie is abundant in multimodal constructions of conceptual metaphors that describe a person's inner, psychological states, and not only a chronological narrative that could be observed from a detached third person view. To put it briefly, the narrative of the movie itself is multimodal. The movie's load of psychological aspects is here analysed without use of psychological or psychoanalytic assumptions. This movie in particular serves well for this task, as its narrative consists in the development of the main character's thoughts and inner feelings in regard to himself via a multimodal conceptual metaphor of "*submarine*". As such, its analysis reveals insights to how the director depicts inner feelings by outer-subjective images, which collapses altogether the inner/outer dichotomy.

7.2 *Submarine's* Subjectively Related Narrative

The movie *Submarine* was released in 2010 and was directed by Richard Ayoade. It is based on the novel with the same title by Dunthorne (2008). That the movie reproduces the novel's narrative is a significant aspect for the present analysis, as it involves a translation and interpretation from a prevalently monomodal medium to cinematic multimodality. Such a translation between different modes is a process of transduction (Bezemer and Kress: 169) whereby it is insufficient to consider the translation of "words" into "images" but, rather, the moving of "semiotic material" across modes. The ways in which director Ayoade transduces written text into moving pictures, involving colours and chromatic contrasts, music, speech and written text are revealing for the phenomenon of conveying information generally.

The narrative consists in the main character, 15-year-old Oliver Tate, psychologically exploring himself. The main character also orally narrates the movie, which is thus constructed in Oliver's subjective experience of the unfolding events. He is a pupil in a small town in Wales, living with his parents. At the beginning of the movie, Oliver asks himself who is he, announcing this existential theme as the very plot of the movie. Throughout the movie Oliver's thoughts are heard, as spoken by him, alongside the narrative. He expresses the fact that he often sees his life from the outside, as a movie, from a third person perspective. This implies that many frames in the movie can be interpreted as Oliver's projection of events. Thus, subjective shots, as termed in cinema studies (see Buckland 2000: 58–68), accompany a subjective narration. Cognitive semiotics assumes knowledge as subjective and embodied. The cinematic medium uses the embodied point of view to convey characters' knowledge: the audience easily recognises a subjective shot which, by its framing, implies a subjective phenomenal experience. As Oliver himself has a cinematic representation of his phenomenal world, representing his life via the cinematic medium is immediately revealing. One of Buckland's main arguments for semiotics as an adequate framework for film analysis is precisely that it acknowledges how society relies "upon systems of signification to structure, simplify, and organise experience" (2000: 6). Therefore, from representing Oliver's world in Oliver's own system of signification stem some intertextual meaning effects, as the audience is led to assume that many of the movie's frames are meant as Oliver's own representations of the events. Thus, a statement is made that Oliver understands his own life in terms of the media characteristic of his society (e.g. cinema). A relation is inferred, therefore, between social reality and representation. In their social semiotic approach, Bezemer and Kress focus on "the relation between social environments and representation" (2008: 167), because, they argue, the contemporary dominance of multimodal representations has produced a change to the classic notion of text and its principles of composition. For this reason, an appropriate text hermeneutic, comprehensive of the multimodal dynamics of cinema, is required.

The movie's narrative proceeds with Oliver starting his first romantic relation with a fellow pupil, Jordana Bevan. Oliver and Jordana are presented as passing together blissful moments as teenagers in love. Their relation is shaken by some external events: it seems that Oliver's parents' marriage is going through difficulties and that his mother is very close to having an affair and Jordana's mother needs to have surgery because of a brain tumour. As a teenager, yet to discover his identity, Oliver does not manage the situation well and his relation with Jordana is interrupted, leading to Oliver undergoing an anxiety that contributes to his self-discovery. His parents' marriage recovers and Jordana's mother survives, but Oliver finds himself in a deeper distress than ever because of his separation from Jordana.

The last scene of the movie consists in the event that brings Oliver and Jordana back together as a couple. This is preceded by a monologue of Oliver's, conveying that he is beginning to discovering his true identity (or personality). This last scene of the movie makes a full use of the metaphor of 'submarine' as the deep inner life of a human being which cannot be noticed by another person. This metaphor is developed throughout the movie, and its construction unfolds the events constituting the narrative. This narrative is divided in five parts, each part being announced with script over black background and a dramatic change of music. The five parts' titles are: Prologue, Jordana Bevan, Graham Purvis, Show Down, Epilogue.

7.3 Charles Peirce's Multimodal Sign and the Body

Biosemiotics and cognitive semiotics are theories of signification which give a central role to the body. Meaning is a competence of the body. Peirce suggested that meaning belongs to the phenomenology of the body, explaining that children make sense of the world and develop their cognitive capacity by observing and discovering their own body (CP 5.229). Peirce inherited or, rather, recovered the Augustinian concept of sign, through the scholastic vein (see Deely 2009). In *De Doctrina Christiana*, after defining the sign, Augustine offers some examples which are useful for the classification of signs. For the present paper, it is interesting how Augustine explains that facial expressions function as signs for a person's inner states. This might be overlooked as a banal matter. However, it also opens up the perspective to a non-dual account of the human being, from the perspective that the human *Umwelt* itself is multimodal. In Sect. 7.5 of this chapter I illustrate how this simple idea is used in cinema. That the relation between physical body and mental states is semiotic accounts for a holistic understanding of the "signifying body". As such, our multimodal hermeneutics is rooted in our embodied conceptualising. The movie *Submarine* uses this implicit assumption in developing several meaning structures. Of course, we can also lie—we can pretend to smile when we actually experience sadness, with the purpose of misleading our beholders. All the more, the fact that we can lie in this manner proves that meaning is construed to various degrees by our body's morphology and activity. Approaching the human body on a biosemiotic and anti-Cartesian account, Hoffmeyer nevertheless agrees in some critical regards with Descartes. Such is the idea that truth, that which really matters, lies in the depths, and not on the surface:

> The most important and essential aspects cannot be seen or sensed directly, we are repeatedly advised, but must be dug out from their hiding places deep within the depths of things. Deepest within, goes this logic, we will find what we most profoundly feel and know. (Hoffmeyer 2008: 171)

For Hoffmeyer this leads not to avoiding the surface of the body in favour of a search into the immaterial depths of the mind, but to realising the importance of knowledge of the body's exterior membrane, namely the skin. Interpersonal communication and, therefore, personhood generally, originates in the skin, by which "we encounter the world arounds us" (2008: 172). This accounts for films' audiences' effortless recognition of subjective shots: knowledge and our interpersonal interactions are wrapped up in a skin. Every subjective shot, therefore, implies a Universe of Discourse, namely that of the subject's present phenomenality.

Stjernfelt's exploration of Peirce's doctrine of Dicisigns (2014) is justified by a biosemiotic rationale: the development of a doctrine of natural propositions, that is, Subject-Predicate structures of the biological world, independent of language. In this perspective, the Dicisign is a feature of embodiment.

7.4 Multimodal Logic

Peirce's doctrine of Dicisigns underpins a multimodal logic in which cognitive semiotics can be grounded. *Dicisign* or *Dicent Sign* is Peirce's notion of proposition, that is, any sign with a Subject-Predicate structure. It is the second type in the Peirce's Rheme-Dicisign-Argument triad, which describes the sign-relation in relation to its interpretant (conclusion). The subject involved in the Dicisign's construction is an Index, pointing out to what the Predicate (which Peirce also calls *Rheme*) is applied. The Predicate is an Icon in use, as it attributes a quality to the Subject. The icon is the sign type which signifies due to similarity (CP 2.247, CP 2.276, CP 5.73, Stjernfelt 2007: 49) and, as such, it is the only sign which can be used as predicated (CP 2.278, Stjernfelt 2007: 76). As such, the Dicisign (*I am happy*) is the sign which conveys information (via the predicate … *am happy*) about something (the subject, *I*). A particular characteristic of Peirce's Dicisign is that it accounts for predicates being able to satisfy more than one Subject in the same proposition (e.g., the proposition *You and I are happy* contains two subjects and one predicate). Peirce defined this sign type as follows:

> A *Dicent Sign* is a Sign, which, for its Interpretant, is a Sign of actual existence. It cannot, therefore, be an Icon, which affords no ground for an interpretation of it as referring to actual existence. A Dicisign necessarily involves as a part of it, a Rheme, to describe the fact which it is interpreted as indicating. But this is a peculiar kind of Rheme; and while it is essential to the Dicisign, it by no means constitutes it. (CP 2.251)

Dicisigns can serve as premises in arguments (CP 2.253), which constitute the most developed sign type, representing "its Object in its character as Sign" (CP 2.252). An interesting characteristic of this notion of proposition is that it affords a multimodal logic. Stjernfelt considers that "Peirce's semiotic theory of Dicisigns does not tie propositions to human language exclusively, neither in the shape of ordinary language nor of special formalized languages" (2014: 51–52). As for any of the sign types, Peirce does not bound the Dicisign to any particular modality. Its composing parts, the subject (Index) and predicate (icon), can each be embodied as belonging to any modalities. This also underpins, therefore, a multimodal construction of arguments. As such, Peirce's dicisign doctrine is most insightful for film analysis. This sign trichotomy therefore endorses a doctrine of Dicisigns rather than of Rhemes, which are most basic, or of Arguments, which are the most complex. The stress falls on Dicisigns due to the embodied structure of knowledge. Since, from a non-dualistic perspective, perception is a component of knowledge, according to Peirce consciousness "is a sign resulting from inference"

(CP 5.313). In this respect, Stjernfelt (2014) explains that the recognition of subject-predicate structures is a basic feature of biological organisms and their evolution. This justifies the crux of Peirce's non-dualistic theory of knowledge, namely the concept of "perceptual judgment" (see CP 4.539), which replaces Kantian a priori and a posteriori judgements:

> Perception does not begin, *ex nihilo*, with sense data - rather, perception continues the chain of inferences already taking its beginning long before our individual births and long before the appearance of our species. This implies that perception for us as "perceptual judgements" - already logically formed and already containing a "propositional stance", claiming some statement about environment aspects. That is no later work of a higher-level logic or language module or anything of the sort - the whole visual system, already in the retina (contours, surfaces, etc.), is geared to stably extracting general aspects from the visual field. True, respectively we may analytically isolate "percepts" as simpler aspects of the perceptual judgement contributed by sensations - but they never had separate existence outside perceptual judgements. Thus, the very perception-action cycle of organisms possesses a primitive inference system - even down to simple reflexive arc versions of it […]. (Stjernfelt 2014, p. 5)

Perceptual judgements have propositional structure, since Dicisigns are the signs conveying information, which recommends them as signs endowed with truth value:

> By a perceptual judgement, I mean asserting in propositional form what a character of percept directly present to the mind is. (CP 5.54)

Because propositions' signification is bounded to the indexical relation to their subjects, they are signs that have truth value according to the principle of correspondence, grounded in its perceptual judgement. Peirce explained that a proposition is false if it "would conflict with a direct perceptual judgment" (CP 2.327). Peirce's examples of perceptual judgements implicitly reveal the multimodal character of dicisigns:

> When the perceptual chocolate comes into view, my imagination of it is a feeling of such a nature that the percept can be compared with it as to size, shape, the nature of the wrapper, the color, taste, flavor, hardness and grain of what is within. Of course, every expectation is a matter of inference. (CP 2.148)

This explains why the environment hardly ever offers monomodal percepts. Also, the way in which Peirce describes this use of inferences underpins the cognitive semiotic account of self-consciousness and otherness, grounded in what Brandt termed "iconic cognition" (2007: 61). Brandt explains that altruism, at least in human beings, is grounded in the possibility of iconically simulating sensations fictionally (2007: 53), arguably a semiotic theoretical ground for the work of mirror-neurones. This explains our effortless recognition of subjective shots in cinema and the strong impressions that these leave.

Several advantages for film analysis stem from the particularities of this semiotic doctrine of propositions. One of these particularities is the possibility of analysing propositions which are not strictly linguistic:

> [...] it allows for the investigation of a broad range of human Dicisigns which do not involve human language - or which partially involve language. This makes possible the study of how pictures, diagrams, gestures, movies etc. may constitute propositions or participate in propositions - highlighting how non-linguistic signs may facilitate reasoning and appear in speech acts taken in a wider sense, including what could be called picture acts. (Stjernfelt 2014: 52)

Cinema language, in the various ways in which it might be defined, is not that language which we, humans, articulate in speech or writing. Nevertheless, like language, cinema language belongs to social settings, as explained by Buckland, which offer the frame for the arguments developed in a film. This is also a characteristic of the Dicisign doctrine as, on this account, Universes of Discourse are framed by the dialogical interaction of Dicisigns. Stjernfelt clearly states that, besides Dicisigns' stemming from evolutionary biology, Peirce's doctrine "embeds Dicisigns and their development in a social setting, Peirce taking the step from proposition to proposition in thought to be dialogical and to presuppose the knowledge of a Universe of Discourse shared among dialogue participants" (2014: 52). Not only does this account for the social embedding of cinematic productions, but it explains the formation of fictional discourses within the movies themselves:

> This further allows for a plasticity of interpretation of Dicisigns, relative to the Universe of Discourse in which they partake. (2014: 52)

It is embodied presence, in the first place, that evokes a Universe of Discourse. Stjernfelt explains that the implicit linking of Indexes and Icons into Subject-Predicate structures is due to co-localization, that is to say, merely the reality of these components' simultaneous presence (2014, p. 108). The Dicisign's quality of conveying about something presently indexed solves the issue of deixis in cinema, in a manner fitting Buckland's theory. Buckland argues that traditional film theories, such as in the case of Christian Metz and Francesco Casetti, do not offer a theory of deixis that fully accounts for the transformation implied in the production process, starting from the filmic text (2000, pp. 52–53). The issue that Buckland finds with these theories stands in their linguistic conceptualisation of deixis, implying that deixis occurs through linguistic pronouns. Instead, Buckland argues for "'deixis' as a broad cognitive concept, rather than a narrow one (which equates deixis with pronouns)" (pp. 52–53). He explains that the implication of freeing the concept of deixis from a purely linguistic construal is that "we can use it to analyse how other audio-visual media are oriented to their contexts of production and reception" (p. 53).

The notion of Dicisign, as the expression of perceptual judgments which allows for the formation of Arguments, contains an Index, as its Subject part. Deixis is implicit in the Dicisign: it is its Subject. Even more interesting and insightful is the fact that the Dicisign is possible due to co-localization, which thus appears as a metaphysical property of reality. The co-presence of two otherwise distinct entities inevitably leads to a perceptual judgment. The Dicisign is self-reflexive by its syntax, which constitutes the starting point of any Universe of Discourse. Thus, for example, when the picture acts of a movie are accompanied by sounds, be it speech or otherwise, we are right to assume that they are connected. Moreover, they are not just connected in incomprehensible ways, but, primarily, in a Dicisign, that is Subject-Predicate, syntax. As such, by seeing the movements of someone's lips and hearing uttered speech, prior to judging the semantic content of what is being said, the audience knows that the represented person's voice sounds in such a way as it is heard.

To illustrate the relevance of Peirce's doctrine of Dicisigns for cognitive semiotics of film in the next section of this paper I explain the meaning effects of some cinematic techniques employed in *Submarine.*

7.5 Camera Techniques in Dicisigns

The movie begins by Oliver Tate's introducing himself and, thus, the main plot of the movie. His voice is heard explaining:

> Most people think of themselves as individuals, that there's no-one on the planet like them. This thought motivates them to get out of bed, eat food, and walk around like nothing's wrong. My name is Oliver Tate.

Just before these words are spoken, the audience is given a view of Oliver's room. The room is heavily personalized by Oliver with a large number of objects (photographs, posters, furniture, books). When the speech begins, Oliver is seen in the frame, which creates several Dicisigns, by the principle of co-localization: "This is the speaker's room", "This young boy in the frame is the speaker." Oliver's lips are not moving, which implies that these are his thoughts. As the words "My name is Oliver Tate" are heard the camera closes-up on Oliver, confirming the mentioned Dicisigns. After a close-up on Oliver's face, the boy moves his eye gaze, until this point fixated towards the window of the room, towards the camera: Oliver is looking at his audience. Thus, deixis can be assumed as on account of co-localization. The film's Universe of Discourse is being framed through this multimodal presentation of Oliver. By these Dicisigns it is conveyed to the audience that the movie offers Oliver's account: the Universe of Discourse is Oliver's phenomenal account of the events to follow. Just before the speech begins, the camera passes by a decorative skeleton standing in Oliver's room. This is telling for the Universe of Discourse, introducing the existential theme of Oliver's speech. Also, it links existential questions with matters of anatomy and biology, as the film suggests in a number of occasions.

The main argument of the movie is also implied visually in this scene, but not conveyed by Oliver. As Oliver is looking at the window, he is sitting next to a big drawing of a submarine. Thus, on account of co-localization, the visual structure Oliver-Submarine is interpreted as a Subject-Predicate structure as "Oliver is submarine." This is the metaphor, announced also by the movie's title, that the narrative, and Oliver through it, explores. In Peirce's taxonomy, the metaphor is a particular type of icon, namely a hypoicon (pure icon) that signifies due to parallelism to something else than the sign:

> Hypoicons […] which represent the representative character of a representamen by representing a parallelism in something else, are *metaphors*. (CP 2.277)

Peirce's account of metaphor is compatible with the general understanding of metaphor in cognitive semantics and cognitive semiotics, as a parallel between two distinct domains of meaning.

At a close look, simple iconic features can be noticed, confirming that it is not an accidental coincidence that Oliver is sitting next to the submarine drawing. The drawing conveys that the submarine is indeed underwater, depicting the level of the surface of the water, and its periscope sticking above the surface. In a like manner, Oliver's forehead is standing above the level of the water depicted in the drawing and his eyes are looking out the window. Thus, Oliver's pose is iconic to the submarine depicted: the eyes are like the periscope, while his life proceeds underneath the water. These iconic relations form the predicate (… *is (like) submarine*) of the Dicisign (*Oliver is (like) submarine*). Thus, the metaphor between the psychological source domain and the underwater physicality target domain is set from the beginning of the film.

A few seconds later the word "Prologue" over black background appears on the screen. This is followed by a scene from a classroom where the teacher's voice is heard to say to the pupils:

> What kind of young person am I? That's the challenge.

In this way the Universe of Discourse is entrenched by a multimodal Dicisign that can be expressed as "The prologue is 'what kind of person am I?'." Of course, the multimodal cinematic language is, in this instance, more adequate than the plain linguistic expression. Linguistically, the Dicisign is rather properly expressed as "If the prologue itself begins with the words 'what kind of person am I' then the film is about 'what kind of person am I'?" By now the audience knows that the film has its own subjective *I*, namely young Oliver Tate. As such, Dicisigns from the previous scene come into semiosis with Dicisigns from the current scene, resulting in the interpretant "The film is about Oliver Tate asking himself what kind of person is he." Also, irony is conveyed through some Dicisigns implied by the way in which words are orally pronounced. Oliver's introductory speech is not ironic, as it feels genuine. Even though his lips are not moving, the audience assumes that the boy in the frame is the author of the words heard. Co-localization is confirmed by the probability that that boy's voice sounds like the voice that is being heard. Therefore, there is at once an indexical and an iconic relation between the boy seen and the voice heard: it is likely that a young boy's voice and discourse sound as such. The teacher's voice is too mature and presumptuous to actually mean the question of "What kind of young person am I?" The voice is obviously not that of a young person. While it is implicitly assumed that the teacher is inviting the pupils to ask themselves the question, the lack of correspondence between voice and discourse challenges the principle of co-localization, leading to a meaning effect such as "This cannot be taken seriously." Given that the Universe of Discourse and several deictic markers (*I* = Oliver) are set by this point, the audience is informed that a pupil such as Oliver cannot make much use of teachers' inputs. This Argument is constructed by a series of Dicisigns: "Oliver inquires about his personality", "Oliver is a submarine", which means that "Oliver may choose not reveal himself to others", leading to the conclusion that "The teacher cannot be taken seriously." An short dialogue between a pupil and the teacher confirms this. After his speech about discovering oneself, the teacher addresses a pupil and is answered back:

Teacher: What do I mean by "self-discovery"? Mark Pritchard?

Mark Pritchard: Having a wank, sir?

The teacher answers by snapping his fingers, pointing towards the door so as the pupil Mark Pritchard understands he is being sent out of the classroom, and continuing his discourse: "Discovery of the self. Who am I?" While it is common sense to assume that the student wanted to be ironic and daring towards his teacher, this answer also serves the larger scheme of the narrative of the film. It confirms the previous Interpretant implied, namely that the teacher does not answer to the pupils' real needs and challenges. Moreover, it reveals the teacher's inability to accept that perhaps masturbation might be an important aspect of the young boys' self-discovery. As mentioned above, the movie suggests on many occasions a correlation between psychological introspection and biology and/or anatomy or human physiology. From the teacher's short dialogue with Mark the statement in the indicative mood "Self-discovery means having a wank" is derived. The indicative statement stemming from the teacher sending Mark off is "The teacher does not take Mark Pritchard seriously." Together with the previously constructed Dicisign, "The teacher cannot be taken seriously", these form an Argument:

"The teacher cannot be taken seriously."

"Mark Pritchard states that self-discovery means having a wank."

"The teacher does not take Mark Pritchard seriously."

Therefore

"Self-discovery is about having a wank."

Thus, another psychology-biology correlation is suggested. This builds up to the framing of the Universe of Discourse, that is, Oliver's own thinking. It is Oliver's own inner world that constructs the submarine metaphor, which makes use of this correlation. While at times the referent is the lifeless submarine boat, the image of *submarine* is used to refer to underwater life. This discourse, I explain below, is constructed via Oliver's father's presence, who is a marine biologist researcher.

Ayoade uses many stills and close-up shots of the characters' faces. This technique was first used accompanying Oliver's statement, "My name is Oliver Tate." By using it in this way, the director not only introduces Oliver Tate, but also accustoms the audience with the technique, as it involves a Dicisign such as "Seeing a face means I am introduced to the character." I argue that this technique is employed for the purpose of conveying characters' thoughts and psychological moods by observing their facial expressions.

For instance, early in the movie, as the character Oliver introduces the audience to his everyday life as a pupil, in the narration of an episode of school bullying, several stills are used to highlight the characters' personalities. Oliver's schoolmate, Zoe Preece, is presented as a usual victim of bullying. On this certain occasion at least, Oliver accepts to bully Zoe in order to please Jordana Bevan. As the bullying takes place, at 07:01 of running time, there is a still on Jordana who is laughing, accompanied by the background narrating voice of Oliver informing the audience that "This [bullying] was the right decision." A multimodal Dicisign is thus implied, which can be expressed linguistically as: "Oliver wants to please Jordana." By this Dicisign, the audience understands that, for Oliver, making Jordana laugh is important enough to overstep some moral principles and become involved in bullying. In his own speech, Oliver uses the word "adapt" to justify his behavior. The use of a term stemming from evolutionary biology to refer to behavior is another suggestion of the psychology-biology correlation. Of course, the Dicisign "Oliver wants to please Jordana" is constructed only based on the more basic Dicisign implied by Jordana's smile, namely "Jordana enjoys bullying Zoe."

A few seconds later another still is used, capturing Oliver throwing Zoe's bag, which he and his colleagues stole and use to anger Zoe, and Zoe's distressed and angry face. Oliver and Zoe's facial expressions exposed here form, as well, Dicisigns revealing psychological moods: Oliver is seriously involved in bullying and Zoe is angry. At 07:33 there is a still on Zoe's face, just as, as a result of the bullying, she is about to fall in a pond of water. Just before this still, Oliver's voice is narrating that "I knew that these events—although hard to bear this precise moment—would help build Zoe Preece's character in the long term." The word *character* together with Zoe's expression of surprise and fear, as she is about to fall, form a Dicisign that revealingly frames the Universe of Discourse: this discourse is about characters. Also, it conveys some Dicisigns about Zoe's character, as perceived by Oliver: "It is in Zoe's character to be afraid", "Zoe has an unstable character", as she physically just lost her balance and is about to fall in a pond.

Close-up shots in combination with jump cuts are used to convey the relation between Oliver and Jordana, before this relation becomes romantic. After the Prologue introduces the theme of self-discovery, the part of the film entitled "Jordana Bevan" follows. Oliver describes in words his attraction for Jordana, comprising aspects of proper erotic attraction, personal admiration and also detached arguments, such as the fact that being seen with her would be good for his popularity. As Oliver explains these in speech, the subject of the speech, Jordana, is shown in various circumstances. Among other situations, Jordana and Oliver are shown passing by each other on a school corridor. As they pass by, they look at each other. As they look at each other jump cuts of close-ups on their faces alternate quickly (06:12–06:15). The close-ups make the facial expressions obvious. Given the characters' facial expressions, these frames produce simple dicisigns that explain the alloscopy between the two:

"Jordana looks at Oliver with indifference."

"Oliver looks at Jordana with caution."

"Jordana sees Oliver like this (cautious)."

"Oliver sees Jordana like this (indifferent)."

Several times characters are shown against a dark background, which implicitly entrenches the viewer's focus on the portrayed character. This happens both in the film's chronological narrative but, also, in a sub-Universe of Discourse of the Universe of Discourse, which consists in Oliver's imagination. For instance, when Oliver and Jordana have a difficult conversation about their problems outside their relation, they are standing in the dark, by a burning container (50:57–52:03). As the two discuss they are filmed against the dark sky, with small and bright city lights in the background. Their faces are illuminated by the fire of the container. Thus, the conversation unfolds in the context set by the dicisign: "They are in darkness." This is a contribution to the "Light/darkness is life/death" metaphor used throughout the movie, which is discussed in the following section of the paper. This conversation is the starting point of the breakdown of the couple. This is conveyed by the frames in which the conversation is presented. In this scene, Oliver and Jordana's faces do not appear together in the same frame. They are filmed either together, but from behind and against the fire which is facing them, or in point-of-view shots, each from the other's perspective. Thus, another Dicisign is formed which conveys what is happening to them, namely "Each is isolated, in the dark." This visual Dicisign confirms what occurs from the dialogue: namely that each of them is concerned about conveying their own problem to the other, but they are not attuned to listening.

Very interesting are the instances of close-ups wherein characters are portrayed against a purely black background. That in real life it is very unlikely, if not impossible, to find ourselves in a purely black surrounding implies that these are not happenings in the time-space in which the film is set. They are sub-fictional instances of the fictional narrative. For instance, at 36:23–36:32 of playing time, individual, separated close-ups of Oliver's mother and father and Jordana, against an entirely black background, accompany Oliver's questions:

What's inside Mum?

What's inside Dad?

What's inside Jordana?

This is part of Oliver's discourse regarding that it is impossible to know the inner life of another human being. His discourse is constructed on top of the *submarine* metaphor, comparing people with submerged objects:

> Ultrasound is a sound vibration too high frequency to be audible. It was first developed to locate submerged objects - submarines, depth charges, Atlantis and such. Some animals, like bats, dolphins and dogs can hear within the ultrasonic frequency. But no human can. No-one can truly know what anyone thinks or feels. What's inside Mum? What's inside Dad? What's inside Jordana We're all travelling under the radar, undetected. And no-one can do a thing about it.

While Oliver's spoken discourse suggests that physical appearance does not reveal what a person truly thinks or feels, I consider that the overall multimodal discourse has a different implication. The multimodal construction comprising the words "What's inside Mum?" accompanied by the close-up on Oliver's mother, against the black background, involves a Dicisign having the question itself as subject and Oliver's mother's portrayal as predicate. This is to say that "Inside Mum is what can be seen." It does not imply that "Everything that there is inside Mum is what can be seen," but only that "This (which we see) is (part of) what is inside Mum." This is confirmed by these three portrayed characters' different portrayals. While each of them is given a close-up against black background, each has a notably very different expression and attitude. While the camera zooms-in on Oliver's mother's face she is staring astray the camera, but when the zoom-in stops in a close-up, she redirects her gaze straight on to the camera. In the case of Oliver's father, he looks away from the camera, detached, throughout the entire zoom-in. Jordana, during the zoom-in, stares straight into the camera. These attitudes are Icons of Oliver's perception of these characters. They are used as predicates in resulting Dicisigns such as:

"Mum minds her own business but pays attention to me when I get close."

"Dad can't be bothered."

"Jordana stares straight at me."

Oliver's above mentioned speech does not imply only that immediate, physical appearance does not reveal a person's inner life. Rather, being "under the radar" implies that a person can choose to remain hidden even from a careful inquiry. As such, what cannot be hidden, that which, so to say, cannot lie, is the body's immediate presence, which proves to reveal all that can be known for certain. By a multimodal construction of Dicisigns it is conveyed that this is Oliver's opinion. The cinematic technique employed here makes use of the very reality that bodily presence implies deixis. This is accounted for in view of Hoffmeyer's concept of the biosemiotic body, as explained above, in Sect. 7.3 of the present chapter.

7.6 Light/Darkness Metaphor

The *submarine* metaphor is developed also alongside a *darkness/light* metaphor. Darkness and light are associated throughout the movie, on the one hand with an individual's inner life, and, respectively, with the outer, social life and, on the other hand with death and, respectively, life. As such, another metaphor results, namely that between isolated inner life/social life and death/life. This conveys that in his psychological formation and search for identity, Oliver discovers that, while he is a 'submarine,' self-isolation can be his death. The metaphoric association of darkness and light with death and life is commonly found in artistic expressions and it has been explained in a cognitive semiotic perspective. Brandt and Brandt, commenting on Edna St Vincent Millay's use of the '*candle burning at both ends*' idiom in the short text 'First Fig', note that:

> The metaphor has an underlying concept – IFE/DEATH IS LIGHT/DARKNESS – but the conceptual metaphor does not account for the possibilities of phrasing and developing meaning offered by the metaphoric expression itself […]. (2005, p. 118)

Brandt and Brandt explain that the meanings of words can challenge physical possibilities so as to make an argument, as it is the case in Millay's lyrics. In the *Submarine* movie, a multimodal medium unlike the rather monomodal plain text poetry, Ayoade uses images of fire, which emits light, and darkness. Such images follow the events and dialogue of the narrative in key moments, so as to construct the metaphor. This is conveyed, of course, by the implied construction of the Dicisigns which participate in the formation of the Argument.

As Oliver is introducing Jordana to the audience, explaining his attraction for her, Oliver is shown to secretly watch Jordana in her moments of solitude. As Oliver is watching her with interest, Jordana, on her own is lighting matches, one after another, which she throws to the ground once their flame is extinguished. This explains Oliver's interest for Jordana: as a submarine, dwelling in the dark depths of his inner self, he is attracted by Jordana's fire (05:52–06:05). Jordan lighting matches is understood in the ground of a basic Dicisign: "Jordana is (or has) fire."After their relationship becomes romantic, Oliver and Jordana light matches together, as a pastime. Jordana enjoys using matches to burn Oliver's hair on his legs. The Dicisigns thus conveyed are: "Jordana burns Oliver," "Jordana can hurt Oliver," "Oliver accepts to be burnt by Jordana." Of course, these are copulated with the general light/darkness metaphor, into "Jordana's fire can hurt Oliver's darkness." In a difficult moment, when Oliver is comforting Jordana regarding her mother's threatening cancer, Oliver offers as a gift to Jordana a pack of matches which he mentions to be her "favourite" (59:27–59:30). The brand of the matches is called "Torpedo" and the pack has a submarine drawn on it. This suggests the complementarity of Oliver and Jordana. Earlier in the movie, Oliver describes himself as a "torpedo," the weapon of submarines, explaining that, to school, he "sneaks unnoticed" (10:55–10:59). Thus, several Dicisigns are suggested by Oliver's offering Jordana the matches, combining in a multimodal Argument rather difficult to convey linguistically: "Oliver gives his darkness to Jordana's fire."

7.7 The '*Submarine*' Multimodal Metaphor

The movie develops around a metaphor of immersion. The semantic field of 'submarine' ('life in the deeps', 'living under the radar') is used to describe a human person's immersion within her own feelings and thoughts. A semiotic web of the 'submarine' is constructed through a discourse comprising articulated language, images and motion, as explained. One of the landmarks around which the metaphor is developed is Oliver's father, Lloyd Tate, who is a marine biologist.

By the various mentioning of themes related to theory of evolution, the movie suggests that Oliver faces the thought of inheriting his father's character. Oliver's faith in classic Darwinism stems more explicitly in the *Submarine* book (Dunthorne 2008) than in the movie, which reveals that Ayoade chose to implicitly frame the theme of natural evolution rather than to make it explicit. This has the desired meaning effect, because of the implicit deixis: the viewer knows that what the movie reveals is the phenomenal world of Oliver. Thus, if natural evolution is framed, then it is framed within Oliver's mind.

The Tate family has a large aquarium of fish in their home, by the kitchen table. Thus, when they are presented as sitting at the table they are filmed against the aquarium. On many occasions Oliver's father is filmed as framed by the aquarium and in some key moments Oliver is portrayed in the same way. This scene implies the Dicisigns "Oliver's father is under water" or "Oliver is underwater." In this manner the characters' psychological moods are conveyed. Also, water, as a physical object, is often present in the film, as Oliver likes to have walks by the beach.

When Oliver is afraid that his mother might be having an affair with Graham Purvis, a former partner of hers, he compares Lloyd's and Graham's discourses. While Graham tends to discuss light and colours, Lloyd tends to discuss the deepness of the ocean and underwater life. This, according to Oliver, makes Graham attractive and Lloyd unattractive. Thus, in view of the light/darkness and the *submarine* metaphors, Graham appears as a lively and social person while Lloyd appears closed within himself and lonely. This is so at least in the case of Lloyd, who is visibly affected by lethargy. Lloyd and Graham are portrayed as antithetical in every sense. While Graham is a researcher and his discourse about the deep ocean is scientific, Graham is a self-proclaimed mystic whose discourse about light and colours is not only unscientific, but empirically not verifiable and manipulative. He uses this discourse to make his living from selling recordings of himself and organising events by which he claims to offer psychological advice. As Oliver begins to be suspicious about the relation between his mother, Jill, and Graham, Jill attends one of Graham's events, where he talks about light. Oliver witnesses the event as well, as he is spying on his mother, unnoticed. Graham's spoken discourse is as follows:

> What is light? What is light? How important is light? OK, let's think about that for a minute. 'Coz it seems very basic and banal but it's not. It's loaded. It's a bloody nail bomb. I think light is probably the most important gift we have from the universe. And if you asked any intelligent bloody writer, like Professor Hawking or anybody of that ilk, he could talk to you… for about, I don't know, a year… just about bloody light! We don't want to be in the dark, being fiddled with. That's for sure.

As he talks about light, Graham makes use of plenty of non-verbal gestures and dramatic changes of intonation. Oliver's reaction, expressed orally, as thoughts to himself, is:

> How's my dad meant to compete? The only time I saw him enthral a room was when he gave a talk about the disappearance of a giant sturgeon.

This statement introduces the sub-Universe of Discourse of Oliver's inner thoughts and imagination and the viewer thus knows that the now unfolding events are Oliver's recollection of his father's talk. In contrast, Lloyd's speech is precise while Graham's is divertive and vague.

It is most telling that in Lloyd's discourse there is mention of the absence of light in the deepness of the ocean. As Lloyd delivers his discourse, he receives from a pupil the question "How deep is the ocean?" (46:42–46:43). This already implies the Dicisign that "Lloyd inspires and answers questions about the deepness of the ocean." His answer to the question is:

> Seven miles at its deepest point, but on average, six miles as a rule of thumb. But no fish could live down there. There's no light, of course, and the pressure is far too great. Any human who found themselves at that level, well, they would die immediately. They would implode. They wouldn't stand a chance. But in answer to your question, the ocean is six miles deep. (46:44–47:07)

Thus, Lloyd speaks about the absence of light, the deepness of the ocean and how, at such a depth, a human would die by imploding. Given the parallelism developed between underwater life and the psychological depths of the human individual, this discourse implies that Lloyd speaks about his own lethargy: there is a level of inner immersion and isolation at which the human individual implodes with his own thoughts and feelings. That there is no light at such depth implies that there is no life. Thus the light/darkness metaphor feeds in the *submarine* metaphor.

Immediately after Lloyd's talk about the deepness of the ocean, the plane is switched to Graham's discourse about light, who is show saying to his audience:

> I am a prism. That's not mad. OK? I am a prism, I am light. I am lucid. I am exciting and delicious. (47:09–47:25)

Thus, these two characters, Graham and Lloyd, are used as the two poles of the "Light/darkness is life/death" metaphor. Graham declares that he is light, implying this about human individuals generally. In contrast, as Oliver is listening to Graham's discourse about how people are prisms and radiate colours he reflects that: "If my dad radiated a colour, it'd be ochre or eggshell" (49:50–49:52). This suggests a Dicisign such as "Dad is a fading colour" or "Dad is colourless".

The change of planes between Lloyd and Graham's speech reveals these claims in the indicative mood, as the viewer notices them one after the other:

> "Lloyd: … the ocean is six miles deep.
>
> Graham: I am a prism."
>
> This results into the Dicisigns:
>
> "Lloyd speaks about the deep ocean.
>
> Graham is a prism."

And these Dicisign lead to an argumentative interpretant such as "Lloyd and Graham are opposites, like darkness and light."

The light/darkness metaphor blends into the *submarine* metaphor in a dream that Oliver has as he is about to go and accompany Jordana at the hospital, in wait for the outcome of Jordana's mother's surgery. Jordana asked Oliver to be present for this and Oliver committed to be there. Nevertheless, after a disturbing dream, Oliver does not go to the hospital. Among other elements of Oliver's dream, when, in the dream, he is about to encounter Jordana's mother, he dreams of himself above a surface of water. Jump cuts between Jordana's mother's body lying on a stretcher and a dark pit which is pulling all the water just under Oliver's feet (1:03:31–1:03:40) create a simple Dicisign, expressible as "Death is a dark pit of water." Waking up from this dream, Oliver does not find the inner strength to accompany Jordana to the hospital. This inevitably leads to the couple's separation.

The last scene of the movie brings Jordana and Oliver back together. The scene explains both that Oliver understands himself better than as he used to in the beginning of the movie and, also, that Jordana understands him. After Jordana refusing to continue the relation with Oliver, the two accidentally meet on the beach. As Oliver is shown watching the sea and the sunset, his inner monologue is heard:

> I don't know if I've come of age but I'm certainly older now. I feel shrunken, as if there's a tiny ancient Oliver Tate inside me operating the levers of a life-size Oliver-shaped shell. A shell on which a decrepit picture show replays the same handful of images. Every night I come to the same place and wait till the sky catches up with my mood. The pattern is set. This is, no doubt, the end. (01:26:43–01:27:25)

The sky/sea contrast evokes the light/darkness metaphor, by now well developed, and identifies, in an indexical way, the sea with darkness. Thus, via two parallelisms, it is concluded that "Underwater is death." The colour of the sky, as Oliver states, can correspond with his psychological mood, as it gets darker, which is another use of the same metaphor. After this Oliver's reflection, he spots Jordana on the seashore, close to the water, and he runs after her. He tries once more to convince her that they should be a couple, but he does not seem to succeed. The conversation, however, takes a turn when Oliver mentions the depth of the ocean. Their spoken conversation ends with the following lines, which are the last linguistic elements of the film:

> Oliver: Ask me how deep the ocean is.
>
> Jordana: Shut up.
>
> Oliver: Go on, just ask me.
>
> Jordana: Why?
>
> Oliver: Because I know the answer.
>
> Jordana: Oh, do you?
>
> Oliver: Yeah. I do.
>
> Jordana: How deep?
>
> Oliver: I'm not going to say.
>
> Jordana: I'm broken-hearted.
>
> Oliver: The ocean is six miles deep.
>
> Jordana: Good. (01:28:46–01:28:59)

While the conversation takes place Jordana and Oliver are mostly filmed against the sea and the sky. This has a strong deictic effect, claiming that "Jordana and Oliver are in light and darkness." This is to say that they are in the right circumstances to discuss their problems. As a further answer to the conversation, Jordana makes a few steps into the sea. Oliver follows her and the two look at each other. Jordana decides to walk further into the sea and the viewer can notice that Jordana has rubber boots, insulating her feet from the water while Oliver does not. Oliver continues to follow Jordana deeper into the sea. The resulting Dicisigns are:

"Jordana walks into the deep darkness."

"Oliver follows Jordana."

"Jordana has waterproof boots, she will not get wet."

"Oliver has casual clothes, he will get wet."

"Oliver can get hurt."

"Oliver follows Jordana nevertheless."

That Oliver accepts the possibility of getting hurt by Jordana's actions is a reoccurring idea. He refused the possibility of being hurt when he refused to accompany Jordana to the hospital. In this final scene, via this multimodal discourse comprising mainly of spoken dialogue and visual images, Oliver argues that he is willing to undergo pain for Jordana's sake. Seeing that Oliver is following her into the sea, Jordana looks at him and smiles. The smile is understood as "Jordana approves of this," "Jordana likes that Oliver follows," "Oliver amuses Jordana."

By Oliver following Jordana into the sea, in the context of the conversation about the depth of the ocean, the interpretant of the Argument is reached that "Oliver would follow Jordana into the depths." As Jordana realizes this, the two can continue as a couple. No words (monologue or dialouge) are needed to explain this, as the Dicisigns constituting the Argument were multimodality constructed.

7.8 Conclusion

The discussed scenes and meaning structures in the movie *Submarine* account for the relevance of Peirce's doctrine of Dicisigns taken to a cognitive semiotic film theory. The primarily Dicent understanding of filmic scenes in Subject-Predicate structures is a simple pragmatic explanation of deixis in cinema. From this basic interpretation, arguments and metaphors can be further developed without confusing Universes of Discourse. Fitting Peirce's notion of Dicisign in a cognitive semiotic framework involves a pragmatic logic of propositions that presupposes embodiment (but not cognition) as a prerequisite of logic.

References

Barsalou, Lawrence. 1999. Perceptual symbol system. *Behavioural and Brain Sciences* 22: 577–660.
Bezemer, Jeff and Gunther Kress. 2008. Writing in multimodal texts: A social semiotic account of designs for learning. *Written Communication* 25 (2): 166–195.
Brandt, Per Aage. 2005. Mental spaces and cognitive semantics: A critical moment. *Journal of Pragmatics* 37: 1578–1594.
Brandt, Per Aage. 2007. On consciousness and semiosis. *Journal of Cognitive Semiotics* 1: 47–64.
Brandt, Line, and Per Aage Brandt. 2005. Cognitive poetics and imagery. *European Journal of English Studies* 9 (2): 117–130.
Buckland, Warren. 2000. *The cognitive semiotics of film*. Cambridge: Cambridge University Press.
Chandler, Daniel. 2002. *Semiotics: The basics*. London: Routledge.
Deely, John. 1982. *Introducing semiotics. Its history and doctrine*. Bloomington: Indiana University Press.
Deely, John. 2001. *Four ages of understanding: the first postmodern survey of philosophy from ancient times to the turn of the twenty-first century*. Toronto: University of Toronto Press.
Deely, John. 2009. Augustine and Poinsot. The protosemiotic development, volume 1 in the *Postmodernity in Philosophy Poinsot Trilogy: Determining the Standpoint for a Doctrine of Signs*. Scranton, London: University of Scranton Press.
Dunthorne, Joe. 2008. *Submarine*. London: Hamish Hamilton.
Eco, Umberto. 1997. *Kant e l'ornitorinco [Kant and the Platypus]*. Milano: Bompiani.
Hjelmslev, Louis. 1954. *Prolegomena to a theory of language*. Trans. F. J. Whitfield. Madison: Universtiy of Madisson Press.
Hoffmeyer. 2008. The biosemiotic body. *Biosemiotics* 1: 169–190.
Jakobson, Roman. 1990. *On language*. In ed. L. R. Waugh and M. Monville-Burston. Cambridge, MA: Harvard University Press.
Kress, Gunther. 2010. *Multimodality: A social semiotic approach to communication*. London: Routledge.
Kress, Gunther, and Theo van Leeuwen. 2001. *Multimodal discourse: The modes and media of contemporary communication*. London: Arnold.
Martinet, André. 1962. *A functional view of language*. Oxford: Oxford University Press.
Nöth, Winfried (ed.). 2006. *Semiotic bodies, aesthetic embodiments, and cyberbodies*. Kassel: Kassel University Press.
Peirce, Charles S. 1931–1935 and 1958. *The collected papers of charles sanders peirce*, ed. Charles Hartshorne and *Paul Weiss*, vols. I–VI, ed. W. Arthur Burks, vols. VII–VIII. Cambridge MA: Harvard University Press.
Saussure, Ferdinand de. 1959 [1915]. *Course in general linguistics*. New York: McGraw-Hill Book Company.
Stjernfelt, Frederik. 2007. *Diagrammatology: An investigation on the borderlines of phenomenology, ontology and semiotics*. Dordrecht: Springer.
Stjernfelt, Frederik. 2014. *Natural propositions: The actuality of peirce's doctrine of dicisigns*. Boston: Docent Press.

Filmography

Ayoade, Richard. 2010. *Submarine*, 157.

Chapter 8
The 'Multi-mode Transitional Practice' of Storytelling While Work Is Done

Derek Pigrum

> *Storytelling is always the art of repeating stories, and this art is lost when stories are no longer retained. It is lost because there is not more weaving and spinning going on while they are being listened to*
>
> (Benjamin 1936/1973: 91).
>
> *So every day I wove on the great loom,*
> *but every night by torchlight I un-wove it*
>
> (Homer's Odyssey: 1990 Book XIX: 358, lines 174–175).

Abstract The present paper is based on a reading of Walter Benjamin's essay The Storyteller (1936/1973) and the author's extensive experience of storytelling while work is done in the secondary school art classroom; a storytelling linked to the author's theory of 'transitional multi-mode practices', Peirce and Sini's semiotics and the object relations theory of D. W. Winnicott. Contrary to what one would expect, the author has observed that listening to a story while working enhances focal attention and provokes doing, in what Benjamin calls a state of self-forgetfulness that impresses the story deeply upon memory. It is argued that this mode of storytelling is a *deixis* that points beyond the horizon of present meaning to a potentiality that is never fully realized but can be returned to, repeated, revised and reconfigured. In addition to Benjamin the author draws upon Gadamer's notion of the story as *ereignis* or 'event' that is also related to Peirce's notion of the immediate object, the energetic and action and final interpretant. Strawson's notion of the episodic, that avoids seeing stories of life experience as frozen into a narrative thread, is discussed in relation to the author's notion of *Das Gegenwerk*, or the work towards the work that avoids definitive closure, where completion continually opens onto 'beginning anew'. It is suggested that the reflexive project of the self, central to much philosophical and educational thinking, be replaced by a notion of *praxis* embedded in place.

D. Pigrum (✉)
International Institute of Semiotics, Kaunas University of Technology, Kaunas, Lithuania
e-mail: derek.pigrum@ktu.it; derekpigrum@gmail.com

A. Olteanu et al. (eds.), *Meanings & Co.*, Numanities - Arts and Humanities in Progress 6, https://doi.org/10.1007/978-3-319-91986-7_8

8.1 Introduction

The theory of 'Transitional Multi-mode Practices' arose out of doctoral research into the creative drawing practices of a range of expert practitioners, their use of multiple sign modes and the author's own artistic practices and teaching experience (see Pigrum 2001). During the course of the research the author explored and developed a link between these drawing practices and Donald Winnicott's object relations theory. Following Winnicott the *transitional object* enables the child to separate from the mother in the process of individuation. The object the child uses for this purpose belongs neither wholly to the external or internal world but exists in what Winnicott termed, *potential space*, as an in-between zone that mediates between the inner and outer world and is the source of all subsequent creativity and cultural experience (see Winnicott 1971). In the author's subsequent monograph, entitled 'Teaching Creativity: Transitional Multi-Mode Practices', storytelling was identified as one of the subsidiary modes of the meta-mode of 'place' (see Pigrum 2009).

Walter Benjamin's essay, 'The Story Teller' (1936/1973), traces the link between storytelling and art and craft practices of the past and describes how stories were told while people wove or engaged in some other activity. He distinguishes between the reception of a story told and reading a novel, or short story, which he sees as responsible for the demise of storytelling, along with the widespread dissemination of 'information' subject to 'prompt verifiability,' something all-pervasive in our or age of information. Benjamin states, "the more self-forgetful the listener is, the more deeply is what he listens to, impressed upon memory" (Benjamin 1936/1973: 91); a self-forgetting induced by working at the same time as listening. The main argument of the present paper is that storytelling while work is done is a remnant of a vernacular practice that not only induces a state of absorbed, relaxed attention but is imbricated with the listener's experiential horizon. It is argued that the self-forgetting induced by the reception of storytelling while work is done is a mode of transitional multi-mode *praxis*, that de-centres an overt concern with what Giddens has termed, the *reflexive project of the self* (see Giddens 1991).

For more than a decade the author observed that telling the story of the Illiad and Odyssey to younger students, in the context of secondary art education while they worked, far from distracting their doing actually provoked it. Stories told to older students had the same effect, falling very broadly into two main categories: those that had an anecdotal quality concerned with *praxis*, and stories based on the teacher's personal experience, unrelated to the work in hand. The present paper is a deeper exploration of storytelling as a mode of place where this is seen as an intertwining of five strands that also possess the four features of Gadamer's notion of *ereignis*, or event (see Figal 2002). An *ereignis* that is both parallel and preparatory to the selective, transitional response to the 'charged' object. A response that exceeds the signified of the object, setting in motion a passage of indeterminate provisional states of doing and undoing.

The first section of the paper outlines the relation of semiotics to Winnicott's notion of the 'transitional object' and 'Potential Space', central to the author's theory of transitional multi-mode practices. This is followed by a section on storytelling as a subsidiary mode of the meta-mode of place consisting of five interwoven strands. One of these strands is Stukenbrock's 'imagination oriented *deixis*'; storytelling as evoking and populating purely imagined spaces with non-present entities "(…) constantly under construction (…)" (Stukenbrock 2014: 71). This leads into a section that addresses storytelling as what Gadamer termed, *ereignis* or 'charged event' and the relation of this event to Peirce's notion of the dynamic object. The penultimate section discusses Strawson's notion of the diachronic and episodic view of life experiences and the implications this has for storytelling (see Strawson 2008). The final section suggests that the present-day emphasis on self-reflection in the context of education be replaced by a concern with *praxis* embedded in place. The conclusion summarises the main points of the paper and suggests further research directions.

8.2 Semiotics, Transitional Objects and Potential Space

Transitional multi-mode practices are based on a response to the encounter with an object in the external world (…) "and both transforms it and lets himself be transformed by it" (Jemstedt 2000: 125). The capacity to respond to the object has its source in the *potential space* created by the lost object of the illusory unity with the mother, or maternal object. The unobtainable lost object opens up a space in which the signified of the object encountered perpetually escapes us, but is something that will always be of interest to us—or, as Agamben states, "(…) something infinitely capable of substitution (without) ever succeeding in exhausting the nullity of which it is a symbol" (Agamben 1993: 33). Sini also uses the word 'nullity' in relation to the fracture or fissure in the unity that preceded transitional object use. Sini refers to the symbolic or 'sign-relation' as an *irrevocable separation* that creates what he calls a *polar space*, where the pointing-to of *deixis* (see Pigrum 2014) is the 'primary structure of the 'sign-relation', as symbol or *symbolon*. Sini's use of symbol is not to be understood in either the sense of language as symbolic sign, or the iconic sign of the lion as a symbol of strength or scales a symbol of justice. Instead Sini insists on the etymological sense of the ancient Greek work *symballein*, as one part of what was once a wholeness, a fragment, stating that only the fracture of the *symbolon* "*allows us to hark back to unity, that in-itself never was*" (Sini in Carrera 1998: 58 emphasis is mine).

Here we enter a realm where semiotic and object relations theory seem to overlap, in ways that call for a more detailed discussion than there is space for here. However, very briefly, for Sini the 'sign occurs as a response'; for Winnicott the transitional object is the response to an object as a means of separation from what he terms the 'good enough mother'. In Sini's understanding, the event of the sign is the opening up of *polar space* and for Winnicott the transitional object opens up a *potential space*. It is with the transitional object that the sign begins to stand in the place of the mother. According to Sini, the basis of the sign refers to the symbolic relation of fissure. For both, the acquisition of the sign is a displacement, a fissure and the infinite withdrawal of Sini's "harking back to unity, that in itself never was".

This harking back is a response to an indeterminate indexical inference that has a potency that prompts a transitional passage of states towards a closure where, in creative practices, each completed work is one in an endless chain of substitutive objects. It is an encounter with an object as a repetition of difference that seeks continuation, a repeated finding, where finding, or in German *finden*, is the root of *erfinden*, or finding that is multiple and continuous invention. Neither curiosity or knowing have a proper place in our response. Rather the charge of the object is a 'turn', that takes us by surprise, a 'charge' leading us down unforeseen pathways in the form-finding of substitutive objects that avoid definitive closure.

8.3 Storytelling as a Mode of Place

The encounter in the external world with the object generates a form finding where the inter-relationship of the Meta and subsidiary modes, shown below in Fig. 1, sets up a web of pathways and inter-changes in an indeterminate, transitional and provisional interweaving of strands. Strands that can be unraveled and rewoven in pursuit of that something that will always interest us.

The modes of place are: the workplace, the place of the page or other surface of inscription, the place of the story, and the virtual space of the screen. It could be argued that storytelling, because it involves the symbolic or linguistic sign, would be better placed in the Meta-mode of the sign. This illustrates the difficulties in sustaining distinctions between the modes because, to some extent, both the Meta-modes of the 'Sign', 'Place' and the 'Operational' are woven into in each other.[1] There are countless examples of the encounter in the outside world that sets in motion the 'cat's cradle' interweave of the transitional in the material practices of artists and people in other fields of creative activity, past and present.

The British painter Sickert (1860–1942) had a lay figure, or jointed wooden manikin, believed to have belonged to Hogarth, delivered to his studio.

[1]The author is not alone in seeing Winnicott's object relations theory as the source of all subsequent creativity. Peter Rudnytsky has edited a book entitled Transitional Objects and Potential Space: Literary uses of D. W. Winnicott. See also Pigrum with Haynes 2008. A 'Multi-Mode' Model of Transitional Creative Practices: The Creative Writing Context. In The Use of English: The English Association Journal for Teachers of English. 59.3. For mentions of Winnicott and transitional objects see also Sherry 2005. The Second Self: Computers and the Human Spirit. Cambridge Massachusetts: The MIT Press. Turkle has also edited 2008. Falling for Science: Objects in Mind. Cambridge Massachusetts: The MIT Press; 2007. Evocative Objects: Things We Think With. Cambridge Massachusetts: The MIT Press; and 2005. See also Burgin, Victor. 1996 Different Spaces: Place and Memory in Visual Culture. Berkeley: University of California Press.

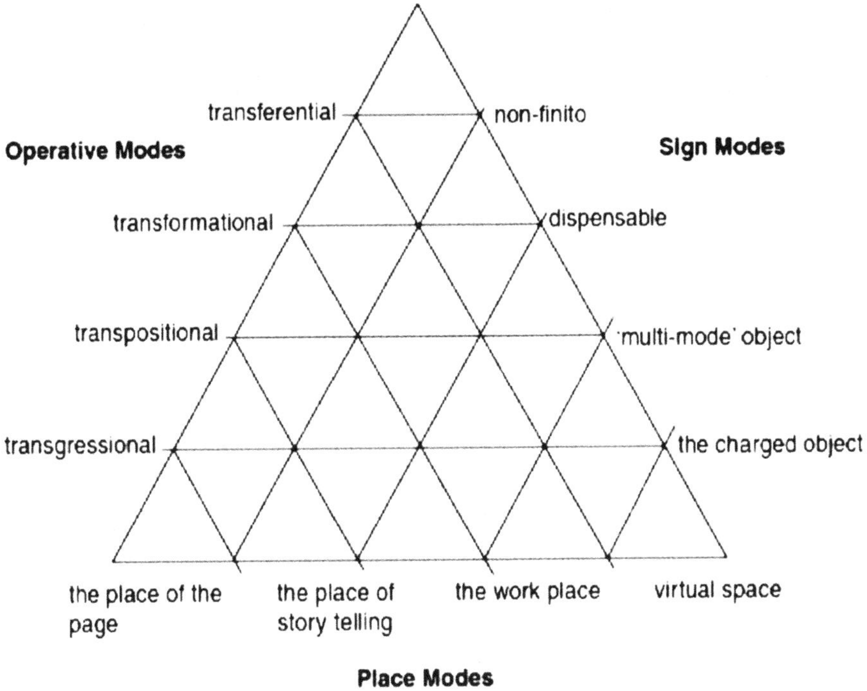

Fig. 8.1 Model of multi-mode transitional practices. Diagram by the author (Pigrum 2009: 10)

The wrapped-up figure had such an immediate charge for Sickert that he painted it directly in chrome green onto the red wallpaper of his studio. This immediate response was followed by photographs taken of the figure that were then squared up and transferred to a painting entitled 'The Raising of Lazarus' (1932).[2] The chance encounter with the object has a profound affect that produces different trajectories of action. It can take the form of a gathering in place that is, at the same time, a way of making it present to the unconscious. We may tarry or linger over it, where tarrying also has the connotation of being slow to act, to postpone, defer or protract, or simply lay aside in the expectation of a building up of different combinations in the unconscious, often in the expectation or anticipation that something will, at some future point, emerge.

However, in terms of the transitional meta-modes, Sickert's response is an example of the immediate response and transfer of the object into a painted image. But, at the same time, we might see the bestowal of the figure as generating an unexpected or transgressional direction. If this work had been developed in dialogue with someone, then it would have been in the order of the dialogic transactional, a way of working common among architects and designers. Sickert's

[2]See Sickert (1992).

knowledge and interest in Christian iconography would make his response a transposition, in the sense of the influence of a precursory image. The influence that Sickert's painting had on the work of Bacon (1909–1992)[3] is also an example of the extremely complex nature of transposition as the influence and response to the work of a precursor, that we will return to later. The passage of preparatory states—in Sickert's case, the squared-up photographs—is in the order of the transformational.

We see storytelling as the weaving together of five spaces: (1) the physical and tangible place of the student's work, the drawing or painting on a surface of inscription, worked on while the story is being told, which also involves inner speech; (2) the 'auditory space' that the fleeting voice of the storyteller creates; (3) the physical place of the classroom and its objects that are introduced into the story; (4) the imagined space generated by the story that displaces both teacher and pupils "(…) in imagination to a place that is either remembered, or constructed on the basis of previous experiences of similar places" (Stukenbrock 2014: 72–73); (5) and, due to understandings, misunderstandings and ambiguities, the story as a *speilraum*, or a space of the speculative play of inner speech.

While it is beyond the scope of the present paper to address Stuckenbrock's paper in any depth, her notion of "imagination-oriented *deixis* (…) instantiated in face-to-face communication, i.e. under the conditions of co-presence and shared perception" (ibid: 71) has significant points of contact with what the author has written concerning storytelling as a mode of place. The story is told by a teacher in the classroom, a story that evokes people and things that are physically absent, creating an *auditory space* that, in its turn, generates an *imagined space* continually under construction and different for each person. Stuckenbrock sees this in terms of a *lamination*, or layering of voice effects, corporeal frames and spaces, both imagined and real. We see this as an interweaving of the voice of the storyteller in the same space as the listener that points to both internal images and associations but also to external objects in the real, rather than imagined, place.

When telling the story to younger students, while they were working on a painting based on symbolic features of the story of Homer's Odyssey, the author would liken the long table around which the students sat to the table lined with axe heads in that part of the story devoted to the competition; or liken the air vent in the ceiling of the classroom to the hatch in the belly of the wooden horse from which, once inside the walls, Odysseus and his companions descended upon Troy. This involved a 'pointing-to' that brings an object into a 'sign-relation' with the story, based on the inference evoked by the words 'imagine this is'. However, to use the words of Bloom, these objects are "potencies, not mere signs" (Bloom 1981: 52). The author's choice of the Odyssey, mentioned above, was not fortuitous. It was inspired in the first instance by listening to the story on Austrian radio told in everyday German speech by Kohlmaier (1949–) and Auerbach's characterisation of the Odyssey as bringing (…) "phenomena together in a continuous and ever flexible

[3]See Daniels (2009).

connection (...)" Auerbach 1968: 6). Storytelling and work is the bringing together of the weave of transitional multi-mode practices with the potency of an event, or in German *ereignis*, in a continuous and ever flexible connection.

8.4 The Story as *Ereignis*

The interwoven strands of the five spaces mentioned above constitute storytelling as an event, or in German *ereignis*, that for Gadamer is: (1) that which happens or takes place; (2) an interaction of elements producing a unity that achieves more than a simple proximity to one another; (3) a mode of disclosure that is taken up and transforms us in some way; (4) an event of meaning that expresses a range of possibilities; (5) a form of disclosure and of appropriation that points beyond the horizon of meaning. The telling of the story holds together in a whole, that is more than a simple proximity of what is said and what is not said but a range of possibilities, the appropriation of which hinges upon the teacher not interpolating her own interpretation of the story but holding it open as a form pointing beyond the horizon of meaning to possibilities never fully realised. Gadamer conceived of the story, in the history of philosophy, as being within the tradition of *praxis*, where storytelling, as an event of meaning, is channelled through an historical consciousness. A consciousness that, because it belongs to a continuous tradition and ritual, encompasses a determinate direction of meaning but, at the same time, eludes a total grasp and is therefore open at each telling to possibilities and the deployment of a new meaning. The assimilation of a story takes time and may never be complete. In this sense the story is always open to the return, the beginning anew, of possessing an ever-present transitional and transformative potential.

There is no binary opposition between the story and the work done but both constitute a wholeness that is beyond the simple proximity of 'multi-tasking'. Through the work done while listening to the story the pupil is in another place. One student reported that in this experience she was only in the place of the story and the place of her work—thus neither wholly in the external or internal world but with a foot in both; in other words in a *potential space*. The speech, that is storytelling, is a sign vehicle that is generated and vanishes in space and time. Painting is also held together, not by what is said and unsaid, but what is beyond saying, the paint that is beyond the horizon of meaning, even when the work itself is figural.

All the varying responses to the event of story or the affective charge of the object are signs that, according to Peirce, would be "(...) a First which stands in ... a triadic relation to a Second, called its Object, as to be capable of determining a Third, called its *Interpretant*" (Peirce in Osborne 2000: 28).

Firstness is then a category of possibility, as a potential that has not yet been actualised and its potency is what Peirce identified as the *energetic interpretant*' or "(...) something which stands to somebody for something in some respect or capacity" (in Osborne 2000: 43) that is in the order of an indeterminate inference that, in the words of Peirce 'evokes an effort' (Peirce 1893). But what is the energetic interpretant? Peirce makes scant mention of the *energetic interpretant* (see The Essential Peirce Vol. 2 1998) but our understanding is that it calls forth an effort of imagination. The story and what Stuckenbrock calls *imagination oriented deixis*, as mentioned earlier, requires the continuous effort of the *energetic interpretant*. In this very important sense storytelling is a preparation for the displacement, or transposition, in imagination of an object or image in the sphere of perception to an absent work in an absent setting. This transposition not only involves the object or image in imagination but also the bodily image of the agent. Thus, in the desire for a given kind of result, we try out something in imagination on a work that is at the present absent.

As we saw above the *firstness* of the lay figure as an energetic interpretant of the Biblical figure of Lazarus combined with the *action interpretant*, that prompted Sickert to paint the wrapped up lay figure directly onto the wallpaper of his studio. But underlying *Firstness*, and both the *energetic* and *action interpretant*, is *praxis*. A *praxis* that prepared Sickert to see the lay figure as a particular sign vehicle with a particular potency and potentiality. In contrast Francis Bacon gathered photographic material that had a charge for him, on the floor of his studio, where it was damaged, smeared, smudged with paint spills and otherwise defaced until, at a particular moment, an image at his feet would emit a potency or charge, and picking it up, and sometimes drawing on it, he would make use of it in his painting. Thus, Bacon waited for things to inadvertently happen to the images that had an initial charge producing a new potency or charge. Sickert's influence on Bacon is, at the same time, an example of the way the work of art never achieves Peirce's conception of a *final interpretant* because they always have the potential to become new *immediate objects*, in a process marked by the mutability of infinite semiosis.

8.5 Against Narrative

In his essay 'Against Narrative' Strawson quotes a number of philosophers, among them Taylor and Ricoeur, who hold that the normative view of narrative as "an inescapable structural requirement of human agency" (Strawson 2008: 196). Strawson identifies the thesis of the descriptive narrative of life experience as a series of connected dots, and the thesis of normative narrative as part of the philosophical notion of living well, that for Strawson is false. What Strawson is seeking, with his rejection of the normative narrative thesis, is to remove us from linking too closely a view of self and the stories of our life experience. With this in mind he sets up a distinction between a Diachronic and an Episodic view of self and experience, where the diachronic outlook sees life as a linear narrative, while the episodic, although it may connect to charged events in the past, does not see life as a linear narrative.

Strawson, as well as describing a range of overlaps between the two psychological outlooks that is beyond the scope of this paper to discuss, mentions a wide range of artists, writers and philosophers that have these different outlooks on life-experience. For example, he identifies Iris Murdoch as having an Episodic outlook. According to Hagberg (2008) Iris Murdoch's diary entry of October 7th 1947 stated that even in the face of the evidence of her diaries, she would "still feel *I didn't know what my past really was*" (in Hagberg 2008: 203 emphasis is mine). Murdoch distinguishes between what she called a 'frozen' and 'unfrozen' past. The frozen past would be what Strawson calls the diachronic where, "one naturally figures oneself, the self or person one now experiences to be, as something that was there in the (further) past and will be there in the (further) future" (Strawson 2008: 210). Episodic experience, on the other hand, is where "one doesn't figure oneself or the person one now experiences oneself to be, as something that was there in the (further) past or the (further) future" (ibid). In other words, in the diachronic approach the past is frozen, its dots connected to form a corresponding, factual narrative thread. That according to Giddens is *the reflexive project of the self* that he distinguishes from the *generic monitoring of action* that, in the following section, we see as synonymous with *praxis*. The episodic is experience as unfrozen and fluid, open to active rethinking. For both Hagberg and Strawson self-understanding does not need to take a linear narrative form but is always a work in progress that the author, in terms of transitional practices, has termed *Das Gegenwerk*—where not only the work, but the picture of the past and the stories of experience itself, avoid definitive closure.[4]

The author's notion of *Das Gegenwerk* is derived from Wuthenow's characterization of Valery's *cahiers* (notebooks) as *Gegenwerk*, in which the two meanings in German of 'toward' and 'against' are implicit (Wuthenow 1997: 158). Earlier we drew an analogy between transitional practices and weaving and un-weaving. The Episodic is the unbinding of the Diachronic signifying process where the weft of remembrance, as a narrative thread, is unpicked in a Penelope-like work of continual beginning anew (Penelope wass the wife of Odysseus, and in his absence avoided choosing one of the suitors that had taken up residence in her house by weaving during the day and then unpicking it at night). Walter Benjamin, in his essay, 'The Image of Proust', speaks of the "weaving of memory, the Penelope work of recollection" (Benjamin (1973: 204). In the episodic approach to the *praxis* of storytelling there is an understanding, however inchoate, of storytelling as *Das Gegenwerk* as an opening onto the perpetual 'towards' that avoids definitive closure.

[4]Pigrum (2007).

8.6 *Praxis* Verses the Narrative of Self-Knowledge

There is an in-built requirement in much present-day education for the pupil to peel off, in layer upon layer, the why and how of their thinking and doing, in the service of *the reflexive project of the self*. In the philosophical tradition of education there is a value commitment to an apparatus of reflective, linear intentionality that pupils are required to live up to as an implicitly good thing; a call for self-reflection that will set their feet on the path to a narrative of self-knowledge. It is a requirement based on the long philosophical tradition that views agency as intentional acts of mind that initiate certain events where the story of these mental representations can be isolated, immobilised and transcribed, in a rational, reflective account. An account that has, as its pre-condition, a withdrawal into the self.

Storytelling while work is done, on the other hand, encourages self-forgetfulness and, at the same time, provokes not only doing, but a wanting to know more about other people's lives than our own. As Silverman states "We look out over the world of visible things and we ourselves are placed in question-not reflexively, not as turning back on ourselves, but as a looking over, around, and through things as they surround us" (Silverman 194: 45). The author sees this as the essence of *praxis*, of storytelling as moving from the archaic periphery to centre stage, as a way of displacing the growing narcissistic concern with self. Stories told while work is done are mostly about the life and experiences of a broad range of people including the painters, sculptors, writers and teachers the author has known. The curiosity, that prompts the call for the story, is a curiosity about other people and their lives.

An example of storytelling and its relation to *praxis* are the stories the author tells to students of the way expert practitioners use dispensable surfaces of inscription, such as an advertising leaflet, a paper plate or the back of an envelope, that underline the provisional nature of the drawing or draft, removing it from the enticement of value, by suspending it in *potential space* or that threshold where nothing is fixed, nothing is static, where everything is provisional and open to the potential that something will stand forth that will activate an alternative path. Very often, in dialogue with students, the author will deliberately use a dispensable surface for them to draw and write upon. In such ways in transitional multi-mode practices, the notion of self-reflection is replaced by *praxis*.

The self is embedded in place and *praxis* with the multiplicity of modes and is de-centred in its response to the object, the event and the precursor. Giddens states, "Only when it is possible to gear regular practices to specifics of place can re-embedding occur in a significant way" (Giddens 1991: 146). A re-embedding made necessary by an overt concern with the reflexive project of the self, with its ingrained philosophical tradition of "(...) whatever we know about the world outside depends on what we can glean from the inner clues" (Hagberg 2008: 225). A tradition of inwardness that neglects the body, place and materiality. Fairfield, in his keynote speech entitled, 'Can We Teach Creativity', delivered at the PSEGB annual conference 2017, seems to relinquish the prospect of creating conditions and provisions in education hospitable to creativity while, at the same time, hinting at what some of these conditions might be—conditions that are very much in line with the author's ideas on *praxis*. While Fairfield touches upon the configuration of private space as playing a role in the conditions hospitable to creativity, he does not conceive of the possibility of embedding these conditions in the place of the classroom, its materials and equipment, as elements in a *potential space* that can induce those states of mind and action that we have termed transitional.

The most valuable advice the author ever received concerning *praxis* was from his art professor Dr. Lilleford RA (1932) who, to allay the author's hesitant first attempts at painting, pointed to the materiality of the surface-a piece of cloth stretched across a wooden frame, the paint as pigment mixed with linseed oil, applied with brushes made of badger's hair. Painting, as provisional as the poet's draft, open to doing and undoing. Advice reminiscent of what Derrida had to say about Freud's 'Mystic Writing Pad' where he insists that, "(…) the wax, resin and celluloid of the Mystic pad, its materiality, emphasise the necessity of the medium of inscription without which there would be no possibility of signification or memory (…)" (Derrida in Johnson 1993: 95). According to Skagestad, in Peirce's *Semeiotic Model of Mind* (Skagestad 2004: 247), artefacts and instruments of inscription, like books, pens, brushes and pencils, exist in a third indeterminate realm between the inside and the outside. As such they are what Winnicott termed a *potential space*, just as the author's former classroom and his present studio are *potential spaces* where the inside and the outside are mediated. It is the author's understanding that *praxis* hinges on the specific, physical characteristics of objects, materials and places to induce certain states of mind and action.

8.7 Conclusion

In an age of modern communications technology and media coverage it would seem that storytelling while work is done is not only archaic, but hopelessly out of step with our times. Why then are we advocating its re-introduction into teaching? One answer lies in our characterisation of storytelling while work is done as a *potential space* or resting place that induces what Benjamin termed 'self-forgetfulness' or a de-centring of the self, a decentring that must come from the outside, from the object, the story or, following Bloom (1981, 1997), the works of a precursor, as the event that is misread, resisted, and reacted against enabling "(…) creation as arising out of the artist's withdrawal from himself" (Bloom 1981: 38), that is at the very core of higher level creativity. The acoustic sign of the story, our response to the 'charge' of the object or the works of a precursor are in the nature of a bestowal, an archaeological find or trace, around which the transformative multi-mode movement of the transitional begins, transforming it and at the same time transforming us.

The archaic practice of storytelling while work is done, opens the horizon of questions that are at the heart of the Western philosophical tradition. Questions of how we orient ourselves in relation to the sign of the object and how this orientation presents the subject with creative possibilities. The doing of the work and the reception of the story involve a concurrence of multiple strands of place and event, or *ereignis*, that induces a self–forgetfulness at odds with the reflective structure of self-knowledge and its concern with the transcendental ego. Following Strawson and Hagberg, we have seen that the diachronic perception of a narrative thread produces a frozen picture of the self, while the episodic approach to life experience is not inert, not frozen, but continually open to beginning anew, closely related to the author's notion of the 'towards' of *Das Gegenwerk.* In the episodic there is an incommensurability where the 'I' can never be fully present to itself. The unpredictable associations the story may evoke in individual reception have parallels in, and are a preparation for, the doing and undoing, gathering and deferral of the charged object as an indeterminate index of agency.

At the beginning of this paper a link was suggested between semiotic and object relations theory. An important future research direction would be to explore the way sign acquisition is conceived of in both object relations theory and semiotics with particular reference to the work of Carlo Sini and Donald Winnicott and the implications this has for transitional multi-mode practices. A further research direction would be to explore, in greater depth than has been possible in the present paper, the relation of the transitional object to Peirce's *dynamic* and *immediate object* and their respective *interpretants*, in order to establish to what extent *infinite semiosis* can be equated with transitional multi-mode practices and the notion of *Das Gegenwerk.* In addition to these two suggestions for further research there are grounds for conceiving, at the very heart of the chain of substitutive objects, a lack, that suggests a relation to notions of desire that has been beyond the scope of the present paper to explore but which would form a third suggestion for further research linked to those mentioned above.

Finally, one of the most pressing tasks of our present time is to keep alive the practice of storytelling as an episodic, vernacular practice, resistant to self-knowledge as a linear narrative and to the present-day demand for the auditing of educational experiences. In the author's experience, storytelling falls outside the managerialism and calculative rationality concerned with the instrumental and quantitative, that along with the dominance of technology has become widespread in schools. The present paper is a contribution to an alternative that Michael Filimowicz and Veronika Tzankova, in their implementation of the author's theory of multi-mode transitional practices, state '(…) is highly adaptable to a variety of learning situations. With a generalizable field of twelve conceptual themes and practical parameters to explore, it is not difficult to find potential mappings and modifications to assignments that enrich the creative output of students (…).' (Filimowicz and Tzankova 2014).

The de-centring of the self of transitional practices is the opposite to the linear notion of, 'I know where I am going because I know where I have been,' built into much of our belief in reflective intentionality that closes the door to the provisional, the indeterminate, to the transitional. This was echoed by the artist William Kentridge in a talk given at the Burg theatre in Vienna about "(…) getting past the good idea (by) allowing the moment to have its place" (Kentridge 2017). The hard and fast intentionality that lives in the 'good idea', is the same as the frozen narrative thread of the past. It must be continually open to an unravelling, a 'beginning anew', that allows something from the periphery to move to centre stage. This is not possible if the diachronic lining up of the dots of lived experience leads to, or at least encourages, the kind of intentionality where one lines up not only the dots of lived experience, but the ducks of intentionality in a linear way, rather than the open, transitional horizon of possibility offered by transitional multi-mode practices.

Acknowledgements The Author acknowledges the small grant from the PSEGB that made it possible for him to present an earlier much shorter version of this paper at the 13th Meeting of the History and Philosophy of Education. June 7–9 2017. School of Education University of Tampere, Finland.

References

Agamben, Giorgio. 1993. *Stanzas: Word and phantasm in western culture. Theory and history of literature*, vol. 69 (trans: Ronald L. Martinez). Inneapolis: University of Minnesota Press.

Auerbach, Erich. 1947/1968. *Mimesis: The representation of reality in Western Literature*. Trans, Williard R. Trask. Princeton, New Jersey: Princeton University Press (in German).

Benjamin, Walter. 1973. *Illuminations*. Trans. Harry Zohn. London: Fontana/Collins.

Bloom, Harold. 1997. *The anxiety of influence: A theory of poetry*. Oxford: Oxford University Press.

Bloom, Harold. 1981. *Bloom: Kabbalah and criticism*. London: Continuum.

Carrera, Alessandro. 1998. Consequences of unlimited semiosis: Carlo Sini's metaphysics of the sign and semiotical hermeneutics. In *Cultural semiosis: Tracing the signifier*, ed. Hugh J Silverman, 48–62. London: Routledge.

Daniels. 2009. Francis Bacon and Walter Sickert: 'Images which unlock other images'. In *Francis Bacon: New studies: centenary essays*, ed. Martin Harrison, 57–88. Göttingen: Steidl.

Figal, Günter. 2002. The doing of the thing itself: Gadamer's hermeneutic ontology of language. In *The Cambridge companion to gadamer*, ed. J. Dostal, 80–102. Cambridge: Cambridge University Press.

Filimowicz, Michael and Veronika Tzankova. 2014. Discursivity and creativity: Implementing Pigrum's multi-mode transitional practices in upper division creative production courses. Presented at ICET 58th world assembly: Moving forward in curriculum, pedagogy and leadership, June 16–19, Canada: University of Ontario Institute of Technology.

Hagberg, Garry L. 2008. *Describing ourselves: Wittgenstein and autobiographical consciousness*. Oxford: Oxford University Press.

Harrison. 2009. Painting, smudging. In *Francis Bacon: New studies: Centenary essays*, ed. Martin Harrison, 143–168. Göttingen: Steidl.

Homer. 1990. *The odyssey*. Trans. Robert Fitzgerald. New York: Vintage Classics.

Jemstedt, Arne. 2000. Potential space-the place of encounter between inner and outer reality. In *International forum pschoanalysis*, vol. 9, 124–131.

Johnson, Christopher. 1993. *System and writing in the philosophy of Jacques Derrida*. Cambridge: Cambridge University Press.

Kentridge, William. 2017. *In defense of the less good idea*. Speech delivered Vienna, the Sigmund Freud Vorlesung in the Burg Theatre 7 May, 2017.

Osborne, Peter. 2000. *Philosophy in cultural theory*. London: Routledge.

Peirce, Charles S. 1893–1913. *The essential peirce: Selected philosophical writings*, vol. 2. Bloomington: Indiana University Press.

Pigrum, Derek. 2007. The 'Ontopology' of the artist's studio as workplace: Researching the artist's studio and the art/design classroom. *Research in Post Compulsory Education* 12: 280–291.

Pigrum, Derek. 2014. Deixis and desire: Transitional notation and semiotic philosophy of education. *Journal of the Philosophy of Education* 48: 574–590.

Pigrum, Derek. 2001. Transitional drawing as a tool for generating, *Developing and Modifying Ideas: Towards a Programme for Education*. Ph.D. thesis, University of bath (Unpublished).

Pigrum, Derek. 2009. *Teaching creativity: Multi-mode' transitional practices*. London: Continuum.

Sickert. 1992. *Paintings*. In ed. W. Baron, O'Connor, A. G. Robins, R. Shone, 294–295. New haven and London: Yale University Press.

Strawson, Galen. 2008. *Real materialism: And other essays*. Oxford: Oxford University Press.

Stuckenbrock, Anja. 2014, Oct. Pointing to an empty space: Deixis am phantasma in face-to-face interaction. *Journal of Pragmatics* 70–93.

Winnicott, Donald W. 1971. *The child and the outside world*. Harmonsworth: Penguin books.

Wuthenow, Ralph-Rainer. 1997. *Paul Valery: zur Einfürung*. Hamburg: Junius Verlag.

Chapter 9
Musical Performer's Corporeal Identity and Its Communicative Functions

Lina Navickaitė-Martinelli

Abstract The present article attempts to discuss several aspects of the corporeal identity of an art music performer, pianist more specifically, as significant carriers of the meanings of the performance itself. The richness of potential significations that a musical performance is capable to communicate is emphasized here, and for that purpose three semiotic models are employed that can contribute to the investigation of corporeal semiosis within the phenomenon of musical performance. Firstly, elements of the inner self and the outward identity, individual and collective subjectivities of the performer's corporeality are outlined within the framework of the semiotic square of performer's subjectivity (after Greimas and Tarasti). In addition, the author's contribution to the already-existing discussion on performer's corporeal identity and communication combines two other celebrated semiotic models: the Peircean icon-index-symbol classification of sign relations, and the model on the functions of language formulated by the Russian linguist Roman Jakobson. Having explained in general terms how these particular theories by Peirce and Jakobson work within the field of performers' corporeality, the suggestion is made to combine the two, for the purposes of encompassing the topic into a single model thus encapsulating different forms of analysis of performers' corporeality.

9.1 Introduction

An observable tendency in Western culture is that of regarding art music as a kind of mental, "abstract" type of human creative activity. Even in the field of music performance, which is essentially related to the performer's technical, or physical, qualities and abilities, the values most commonly referred to are those of spirituality, creativity, or intellectuality, among others. In the mind-body relation, the latter is generally thought of as a mere adjunct to the "higher" realm of the former. However,

L. Navickaitė-Martinelli (✉)
Lithuanian Academy of Music and Theatre, Vilnius, Lithuania
e-mail: linanavickaite.eu@gmail.com

A. Olteanu et al. (eds.), *Meanings & Co.*, Numanities - Arts and Humanities in Progress 6, https://doi.org/10.1007/978-3-319-91986-7_9

a semiotic theory (drawing a lot from theatre studies, and the other way round) has the advantage of not forcing any unwarranted separation between the spiritual and the physical, nor limiting itself only to the cerebral side of musical performance.[1] One of the possible ways to analyze musical performance in semiotic terms, as shown in Barthes (1972), Clarke (1998), or Cumming (2000), is to pay attention to the relation of mind and body in performing music and to acknowledge that a significant part of the performer's identity stems precisely from his/her bodily signs.

Even though general musicological discourses on music performance have for a long time been dominated by a clearly "intellectual" approach, things are changing as Performance Studies and practice-based research become interested in such issues as expressiveness of music performance (Fabian et al. 2014) or gestural communication between chamber musicians (McCaleb 2014). Several studies from the fields of cognitive neuroscience of music (Gaser and Schlaug 2003; Krings et al. 2000; Koeneke et al. 2004), empirical musicology (Clarke and Cook 2004, a.o.), and, more specifically, performance science (starting with Williamon 2004) exist, dealing with various psychosomatic aspects of a performer's work, with physical and psychological elements pertinent to musical performance and relative as well as causal interrelations of these two aspects. Some of them tackle explicitly the visual component of music performance, acknowledging that "performance is not only a sonic event" (Clarke 2004: 92), and judging the expressive properties of various performances on the basis of visual and aural stimuli experienced through live observation and/or video data (e.g. Davidson 1993, 1994; Thompson et al. 2005; Schutz 2008) of concrete performances. Some others concentrate on the ways for enhancing performance and are more applicable to and/or interested in the learning of music and teaching processes. There have been various attempts to classify or otherwise analyze the gestural language of a performer during the performance—both through the process of rehearsing and at a public performance event (Delalande 1988; Wanderley 1999; Elsdon 2006; Virtanen 2007, a.o.). Scholars in the fields of feminist theory and new musicology have obviously theorized how a gendered body of a performer functions as a channel for expressing corporeal meanings of music and the musician (e.g. Butler 1990; McClary 1991; Kopelson 1996; Mitchell 2000, not to mention a vast literature in theatre studies).

In the present article, an attempt is made to discuss a few aspects of the corporeal identity of a performer as important carriers of the meanings of the performance itself. The richness of potential significations that a musical performance is capable to communicate shall be emphasized here, and for that purpose three semiotic models that can contribute to the investigation of corporeal semiosis within the phenomenon of musical performance shall be employed. To begin with, a few aspects of the performer's corporeal identity—the outwardly visual and inner somatic elements of music performance art—shall be outlined in relation to the semiotic square of performer's subjectivity.

---

[1]This article draws on some previous research carried out by the author while preparing her doctoral dissertation, as discussed in Navickaitė-Martinelli (2014).

9.2 Corporeal Identity and Subjectivity of a Musical Performer

> […] I was watching the Bach Concerto for Two Violins on television. I would not have wanted to miss the smile exchanged by the soloists before each of their shared reprises. (Nattiez 1990: 45)

As Naomi Cumming points out, although a listener's attention, when playing a CD, may not be directed to bodily actions, the impression of a "personality" can be gained subliminally through the markers in sound of what seem to be the performer's characteristic physical responses (Cumming 2000: 22). These characteristic bodily responses of a performer are discussed below as determined by several factors: the capabilities and limitations of the performer's individual body as such, and also by the sort of "behavioural codes", the sets of norms and standards that exist in the interpretation schools and concert practices of Western musical performance, and that have a strong impact on them.

In order to illustrate the inner and outer influences, individuality and standards underlying the creative work of a performer, a semiotic model was developed here which combines both the inner self and the outward identity, the performer's individual and collective subjectivities. We speak here of two kinds of signs: endo-signs and exo-signs, inner and outer characteristic features that dominate one's interpretative choices and constitute a semiotic identity of a performer. The scheme in Fig. 9.1 is elaborated from Eero Tarasti's semiotic square[2] of performer's subjectivity and his four logical cases in the light of Hegel and Fontanille[3]. The way I see it, the *Moi*[4] side of the performer's self is intrinsically related to the Greimassian internal (endogenic) modalities Will and Can, while the *Soi* part of one's identity is reflected by the external (exogenic) modalities Know and Must, thus encompassing all the spheres and categories through and in which the performer's art is communicated.

(1) Greimassian modality of "Will"[5] corresponds here to an inner identity (*Moi*) of a performer. This part might include a personal (family) background, artistic

[2]Due to its wide usage and familiarity, here, the semiotic square is employed merely as a schematized illustration, without its further theoretical explications.

[3]For a broader explanation of these philosophical and semiotic concepts cf. Tarasti (2000, 2005, 2012).

[4]Several authors in the field of semiotics have been dealing with the concept of Semiotic self (originally introduced by Thomas A. Sebeok in 1979), which consists usually of two aspects: an inward and an outward side within the subject. Among these dualities, we have, for instance, the "I" (self as such) and "Me" ("I" in the social context), as used by George Herbert Mead; *Moi* and *Soi* by the French authors (Ricoeur, Sartre, Fontanille use these concepts in their writings); Controlling, deeper self versus Critical self by Charles S. Peirce; or the Bergsonian differentiation between the "superficial" and the "deep" ego.

[5]Modalities of Will (*vouloir*), Can (*pouvoir*), Know (*savoir*) and Must (*devoir*), offered by Greimas in the field of linguistics, were for the first time applied in musicology by Eero Tarasti. Modalities provide music with the semantic meaning. In that they can be *endogenic*, i.e. inherent to the

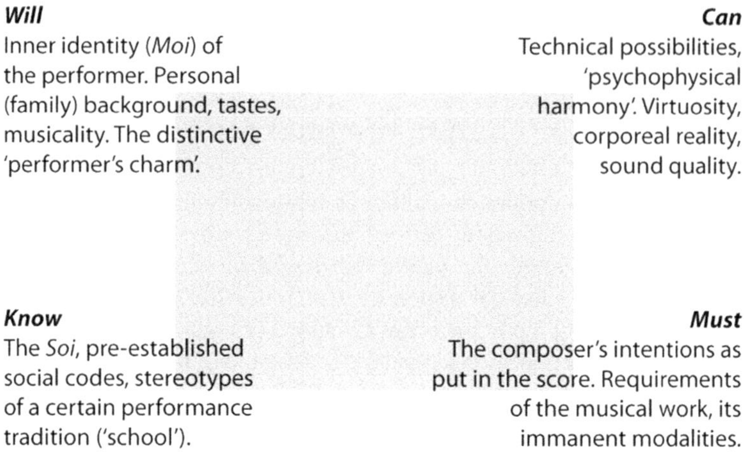

Fig. 9.1 Individuality and standards in the art of a performer (after Eero Tarasti's theory of subjectivity)

tastes or religiousness if any. It also encompasses the person's musicality, together with a distinctive "performer's charm", or charisma.

(2) "Can"[6] embraces the performer's technical capabilities, "psychophysical harmony". Virtuosity, corporeal reality, a certain quality of sound, realized in performance, are also included here.

(3) "Know",[7] or the *Soi* of a performer, consists of pre-established social codes, stereotypes of the given performance tradition ("school"); usually, manners of playing are rated according to certain standard criteria, with some adjudged better than others.

(4) "Must"[8] means the composer's intentions as put in the score, the work's immanent modalities. It is something called out by the requirements of a musical work—the subjective potentialities of its style, without which the work could not "live".

---

immanent meanings of the music, or *exogenic*, that is, "activated" from outside depending on how the music is interpreted and performed. While talking about the composer's work, Tarasti explains the modality of "Will" as follows: *vouloir* appears in, say, Beethoven's sonata in those episodes where the composer is particularly heroic, that is, the way he wants to be (here and below—from Tarasti's Musical Semiotics seminars at the University of Helsinki, year 2005).

[6]"Can" is an inner force, the possibility to express one's Will. In composition, that would be the author's creative power, compositional ideas.

[7]"Know", in composition, means how the composer masters his/her compositional technique, and how well s/he is capable of employing the knowledge (of certain rules, for instance) while composing.

[8]"Must" is the genre-determined and other rules, norms, etc. that the composer must follow in his/her work.

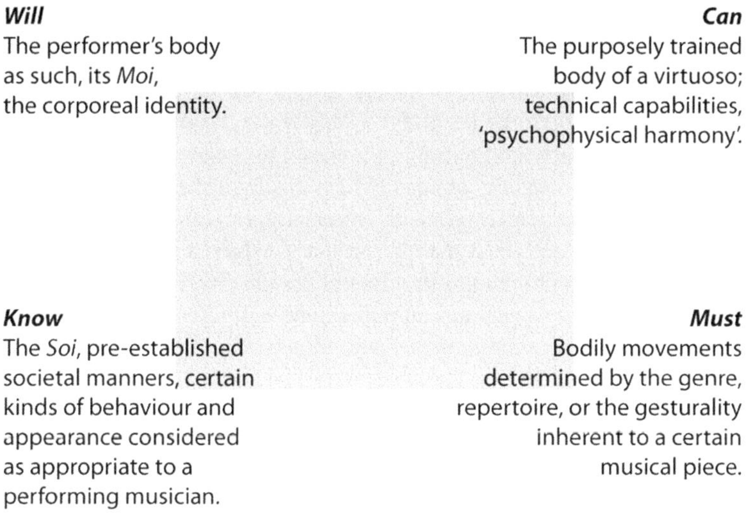

Fig. 9.2 The corporeal identity of a performer, after the semiotic square of performer's subjectivity

With the aim of deepening the theoretical aspect of a semiotic investigation of the corporeal side of musical performance, below in this text I elaborate on the four modalities adapted to our current purpose (see Fig. 9.2). In this case, the model can be described in the following manner:

(1) The Greimassian modality of "Will" denotes a performer's bodily nature as such, it's *Moi*, the corporeal identity.
(2) "Can" is a purposefully trained body of a virtuoso who is seeking to master a certain musical instrument; performer's technical capabilities, "psychophysical harmony".
(3) "Know", or the *Soi* of a performer, consists of pre-established societal manners, certain kinds of physical behaviour that are commonly accepted as pertinent to the performing musician. And, finally,
(4) "Must" means the bodily movements that are determined by the genre, repertoire, by the gesturality inherent to a certain musical piece.

With this in mind, it is important to realize that the corporeal identity of any performing musician encompasses all four quadrants of the square, only some features are more distinct than the others in individual cases. Thus, while attributing any of these qualities to particular artists, these should be understood as merely the most revealing instances, instead of suggesting that some artists possess only one distinctive feature and not the others. Let us start a more detailed discussion from the modality of "Will".

9.2.1 *Performer's Body as the Message of "Will"*

To begin with, the main assumption here is that not only certain messages are conveyed through the performer's body, in this way communicating to the listener or spectator, but also the body itself is a message. Many times we might have noticed that there is a subtle nuance which distinguishes a performance with optimal lightness and grace, or, with concentration and deepness, from one acceptably competent and yet somehow ordinary, which apparently lacks nothing and yet is not entirely convincing. In terms of physicality, it might be just a slight difference of shaping one's gestures, and this minute difference provides a huge one in how this performance communicates and affects the whole process of listening and perceiving. In other words, the apparently insignificant transformations that a certain performer brings to a piece of music using his or her bodily gestures as a means of conveying the message provide this piece with important qualitative changes. Often, performers even intentionally "act" while playing in order to convey to a listener as clearly as possible, with the help of their facial expressions or other bodily movements, the musical significance.[9]

At a very perceptual level, it is a performer's body as such that creates the identity and charisma of an artist, or as Jankélévitch (2003) calls it, the primary *charme*. How else, if not in terms of their personal identity could we define the all-body motion, the singing and conducting of Glenn Gould, the flat and elastic fingers of Vladimir Horowitz, the large hands of Sergey Rachmaninov or Ivo Pogorelich, the singing and standing up of Keith Jarrett, the generally still posture of Radu Lupu, or the purposefully (so it seems) hypnotizing looks at the audience by such artists as Gidon Kremer or Yo-Yo Ma? It is rather improbable that, after once having seen the gestures and facial expressions of Lang Lang, it would be possible to dissociate one's listening to this artist's playing without having in mind this excessively corporeal side of his performances.

Quite obviously, and that would be emphasized by many a practicing musician, certain performative movements depend on the obvious physical requirements of the act of performance itself: e.g. the weight must be delivered to the hand from the spine or the abdomen[10]; one has to achieve a completely relaxed and yet concentrated wrist, etc. And yet, these movements usually differ as related to the particularities of a given body. One performer has larger hands, therefore, s/he is shaping his or her gestures accordingly; another is able to produce a deeper, more rotund sound because of the bigger weight of the arm and hand, and so on.

---

[9]Recent research in performance science and related disciplines has shown how important the facial expressions and bodily movements of musicians are in communicating emotions in music (cf. Thompson and Russo 2007; Livingstone et al. 2009, among others).

[10]For instance, when observing rehearsals or masterclasses by Ivo Pogorelich, a pianist whose tone range is perhaps one of the most remarkable features of his pianism, one sees how many hours may be spent for only achieving the right way to produce (and to *make it last*) one single sound.

Moreover, as the instrument is for a performer some kind of extension of his or her body (as anthropological studies suggest), it could be maintained that the bodily relation with the instrument must be discussed in terms of the *Moi* as well. Performance encompasses a physical relationship of the artist and instrument: a great deal of sensory perception is involved in the act of musicianship, and achieving a good "relationship" with one's instrument is crucially important. It is much easier to obtain such intimacy, say, for the string players due to the fact that their instruments are portable and the player normally plays one or two instruments at a certain period of time, however, some of the greatest piano figures who could afford possessing their personal pianos knew that as well. It has been reported that Josef Hofmann would demand a narrower keyboard; Vladimir Horowitz—a very shallow keyboard that would allow him to play with flat fingers thus achieving a supremely light *pianissimo*; Glenn Gould needed to feel the illusion of vibration while playing—for that reason, one of his own pianos, where the keys were made more distant from each other than usual, was being taken on tour with him. All of these peculiarities were determined, surely, by the corporeal (along with psychological) characteristics of the individual players: what is good for one would be not acceptable for the other. As the Lithuanian pianist Petras Geniušas attests, he cannot imagine that he himself would ever need to feel the physical illusion of vibration, but it seems only natural to him that somebody else (such as Gould) might have needed it. To sum up the importance of the tactile experience, Geniušas' praise of piano as a very *complete* instrument is telling:

> [A piano] is superior to an organ because of its connectedness to the human body. It takes a human body to produce all *crescendi* and *diminuendi*, and the immediate contact with the instrument means worlds. The strings are also related to the human body, there is always that moment of touch, of dynamics, but they are only equipped to produce one, at most two or three sounds at a time. The piano offers the totality of a full score. What a wonder is the piano pedal: it allows you endless possibilities of overtone play, all the riches within it, the bottomless possibilities of mixing and of diluting sounds – the entire alchemy of the piano. (Geniušas, in Navickaitė-Martinelli 2010: 96)

9.2.2 *The Modality of "Can" as Inevitable Vehicle for Professional Communication*

To proceed with the issue of the instrument, we enter the realm of the "Can" modality. Each of them having their individual bodily nature, performers have been taught since the early days of professional training the specific elements of mastering their instrument. Countless scales and exercises are meant for this purpose, not to forget the more difficult tasks of the pedagogical system—to teach one the ways of using the natural weight of arm, the proper way of sitting that would help the body to relax, and so on. Thus, one can say that the body of a performer has to get accustomed to the chosen profession. Surely, some professions are more demanding than the others—a singer is a supreme case, where the body and the instrument are one and the same, the vehicle through which all the visual and sonic significations appear. However, through the constant repetitions of the same movements, also instrumentalists experience rather significant changes of their physical qualities in the course of their careers. Certain learned gestures or movements are repeated over and over again under the same circumstances and in a manner as similar as possible. No doubt, this leads to some kind of "adopting" these gestures, to making oneself master them as if they were the performer's own and, in the successful cases, reaching a sort of "psychophysical harmony". Watching the great piano masters such as Vladimir Horowitz or György Cziffra, or their younger virtuoso followers, perform what fascinates most at a visual level of perception is the apparent ease with which even the extremely difficult pieces are played.[11]

Certainly, not only the virtuosic side of the performance, but also, for instance, producing "a beautiful tone", through a well-balanced physical adjustment to the instrument, is central to creating the impression of musical personality. The "sonic self", borrowing the term of Naomi Cumming, is thus conceived (Cumming 2000: 22–23). However, what counts as the "beautiful", or "natural" character, Cumming adds, is determined by the performance tradition. The performer's body is entrained, or habituated, to produce a certain kind of sound, according to the tradition they follow (Cumming 2000: 29)—this aspect of musicianship, in fact, may well be discussed within the analysis of performance schools.

9.2.3 The Musicianly Role and Its "Knows" Within Society

Whereas the two aforementioned aspects ("Will" and "Can") were related to the inner identity, the *Moi* of a performer, the other two (and in this section we shall be discussing "Know") are from the outer field. As already mentioned above, the *Soi* part of the dichotomy of one's semiotic self consists of pre-established societal manners, certain kinds of behaviour that are accepted as appropriate to the performing musician. Alan P. Merriam writes in his *Anthropology of Music*:

> As there are specific kinds of physical behavior concerned with the manipulation of voice and instruments, there also seem to be characteristic bodily attitudes, postures, and tensions, and it is possible that such bodily characteristics can be correlated with other behavioral elements to reveal significant facts about music making. (Merriam 1964: 108)

[11]Despite the fact that some of the aforementioned ease may indeed look "artificial", it is very interesting how, commonly, the portrayal of the performing musicians in cinema ends up as a failure. Anybody familiar with the process would perhaps agree that performers while playing, first of all, are doing their job and *listening* to what is being conveyed through their bodies. When watching the recording of, say, Wilhelm Kempff playing the *Moonlight* sonata (EMI Classics 2003), what strikes first and foremost from the corporeal side is a very still posture and evident listening to himself. In other words, the performer knows what s/he wants to do, and then tries to achieve it physically. In all the movies on musicians, on the contrary, a bodily portrayal of the "feeling", as intense and apparent as possible, is attempted.

What significant facts can be revealed by the behavioural elements? First of all, we may learn to which culture the performer belongs. As is well known, in some cultures the faces of musicians must remain perfectly impassive, while in Western concert practices we can much more often observe at least some slight expressive changes in performers' eyebrows or other parts of the face; in gospel singing, the chorus is moving in response to the beat, while in the Western art choir music the members of the choir more often remain stable; not to mention hugely different character of bodily motion in popular versus art music culture. Video footage from the La Nit festival at Montjurich Castle in Barcelona (October 8, 1988)[12] of three singles for the album *Barcelona* where Freddie Mercury and Montserrat Caballe sing together remains one of the charming examples of how the two areas merge in terms of physicality, appearance and behavioural codes (from voice formation to dressing to facial expressions) of performers, how does one style follow another, borrow and/or benefit from it, what liberties or restrictions are abandoned in each case, etc.

It would be nearly a truism to state that a musician specializing in Western art music plays in our society a rather specific role. Hence, according to the constraints of this role and his/her rather special status s/he must follow certain consensually accepted codes of what should be the proper behaviour for the musician. Again quoting Merriam, "Musicians behave socially in certain well-defined ways, because they are musicians, and their behavior is shaped both by their own self-image and by the expectations and stereotypes of the musicianly role as seen by society at large" (Merriam 1964: 123). In addition to some sort of 'elitist' manners pre-defined by the status of the art-music performer, there are also purely professional aspects of the matter. Since the beginning of professional training, the future musician knows what the characteristic performance postures are (starting with the very practical thing of the playing itself and heading towards the more socially defined ones) and he or she is taught how to confine oneself to them. There are specific learned ways of how a musician walks onto the stage, how s/he looks at the audience, takes a bow, sits down at the piano or shakes hands to the orchestra musicians.[13] However, even though some of those gestures are rational, cognitive messages, they may affect the listener/spectator on a quite different level than expected.

A specific observation should be made on what might be called a sort of *misbehavior* of a classical musician. As stated above, there is a tendency in Western culture to regard art music, including its performance, as a more mental (as opposed

---

[12]See, e.g. "The golden boy", at https://www.youtube.com/watch?v=OPizyP4x30I (last retrieved: August 2017). The impression is only reinforced by the early demo of the song featuring Mercury's voice only, where he also sings the later-operatic part of Caballe (cf. https://www.youtube.com/watch?v=08qGc43xS10, last retrieved: August 2017).

[13]It is often revealing how one is seen and interpreted from outside, in this case – what is the image of an art music concert ritual in popular culture. *The Cat Concerto* cartoon from 1947 (with the cat Tom, a piano virtuoso, giving a piano recital) is a telling parody of the behavioural clichés and the existing codes of on-stage music making.

to physical) type of human activity. Paradoxically enough, quite an important impact on the listeners' reactions in a concert hall is the bodily posture and behaviour of a performer. "Why do we buy concert tickets when the same sounds can be heard more cheaply and comfortably within our own homes? Why do popular music concerts often include elaborate lighting and staging effects for what is ostensibly an auditory event? Why can't orchestral musicians wear t-shirts and flip-flops?"—asks Schutz (2008: 83), later answering to these rhetorical questions with the emphasis on the importance of visual information on shaping the musical experience and its communicative powers in a seemingly sonic event such as musical performance. Undeniably, the visual aspect of the performance is important to how we perceive the whole act of musicianship. Even taking into account the possible variety of media to experience the performance, once a listener/member of an audience has seen a particular performer onstage, it is hard to believe, as in the aforementioned example of Lang Lang, that this image will be completely absent while listening later to recorded performances of the same artist. There is more than one instance to be exposed concerning the prejudices of listeners towards performers, judging solely by their apparent "misbehavior". In particular this is evident within a special phenomenon of the art of musical performance, namely, competitions. Being the space where a performer by default is supposed to make an effort to be "liked" by the majority, competitions provide multiple examples of rather extreme judgment based on the performer's looks. In a Lithuanian context, perhaps the most famous illustration of this statement that comes to mind is the case of the winner of the M. K. Čiurlionis international piano competition in 1999, a Polish pianist Jan Krzysztof Broja, which suggests that a performer risks having a highly controversial reception only because of his seemingly arrogant posture on stage.[14] Such reactions to performers' "misbehavior" may perhaps be ascribed to the Peircean *firstness*—a person is judged in terms of an immediate impulse; what remains in the receiver's mind of a person, as Tarasti puts it in his *Existential Semiotics*, may be "some striking gesture, perfume, expression, some quality beyond the 'official' image of his personality" (Tarasti 2000: 9), yet one may clearly trace here also rational socially pre-determined expectations. A different case from Broja's, however followed with a similarly aggressive reaction, was that of the Bulgarian pianist Evgeni Bozhanov at the XVI International Chopin Competition in Warsaw in 2010. Bozhanov's onstage facial expressions provoked heated online discussions, and, at times, even openly mocking attitudes towards his playing. This only once again proves that the performers must, in most of the cases, accord with the certain prescribed behavioural patterns, and this links us well to the last of the modalities, namely, "Must".

---

[14]This talented pianist was received with open controversy based not as much on his interpretations (which earned him first prize in the competition) but on his on-stage manners: too "arrogant" entrance (indeed slightly reminiscent of Ivo Pogorelich during the famous Chopin competition), movements and looks *not appropriate* to a classical musician. This situation, and Broja's behaviour in particular, was even discussed on the TV-programme after the competition, where his colleagues (and former contestants) pianists were sharing their opinions on the matter.

9.2.4 The Performer's "Must" as Intrinsically Coded Constraints

This modality manifests itself on several levels of the art of musical performance. First and foremost, this is the level of gesturality as determined by a particular piece of music, the "composed gestures" (discussed at length in Virtanen 2007; also cf. Hatten 2004). Apparently, a performer still has some freedom within certain limits, e.g., in choosing the tempo or the degree of stress in playing a passage, but, as Cumming writes, "To suggest that a performer may decide freely on what melodic patterns are to have 'gestural' connotation is, however, to underestimate the information that may be derived by looking at patterning in scores, or the constraints of style" (Cumming 2000: 136).

Proceeding with the "Must", we also reach the level of a particular style of music, a certain epoch with its stylistic requirements. Some patterns, for instance, in Bach's pieces or in Classicist music certainly require a particular manner of stress and release. At the same time it must be noted that romanticized performance is widely spread in current performance practices, which tends to overemphasize a great deal of musical elements, thus denying the stylistic demands of the previous epochs.

No less than the aforementioned reasons, the boundaries of genre and repertoire determine bodily movements of a performer. To draw but one from multiple possible examples, the great jazz pianist Keith Jarrett is known for acting in a particularly expressive manner while playing: that is, humming himself or producing some kind of screams, constantly moving his head and the whole body, and even standing up while still with his hands on the keyboard. Evidently, when one is playing a Bach prelude or a Beethoven sonata, one is behaving according to the constraints of the appropriate behavioural codes. Although even here some deviations are possible: on some occasions, the bodily movements also reveal the interests of a performer. For instance, after being involved into a few jazz projects and after being given some bodily-related improvisation lessons,[15] Petras Geniušas started using evidently jazzy, swingy movements of the head and body while playing classical music recitals.

Thus, as it has been demonstrated, the variables that determine bodily movements of a performer are of very different types—from the difficulty of a piece or the particularities of performers' hands to the cultural codes. No doubt, certain elements cannot be strictly located, that is, some kind of "grey areas" exist here as well—if we tried to combine the physical side of a performance with the emotional one, certain expressions of the same emotion would definitely belong at the same time to the several quadrants of the aforementioned square. Due to the complexity of the issue, and despite the recently-growing number of studies in the field, the

[15]From personal communication with the pianist, Geniušas 2008.

semiotics of a performer's corporeality as one of the important aspects of performance practices still calls for interesting research directions—some of which are offered below in the present article.

9.3 Corporeality as Communication: Semiotic Models for Analysis and Classification

In addition to the reflection within the framework of the semiotic square of performer's subjectivity that was introduced above, my contribution to the already-existing discussion on performer's corporeal identity and communication combines two other celebrated semiotic models: the Peircean icon-index-symbol classification of sign relations, and the model on the functions of language formulated by the Russian linguist Roman Jakobson.

9.3.1 Peircean Classification of Sign Relations as a Tool for Analyzing Performers' Corporeality

Of all the accounts on musical performance conducted in semiotic terms, the deepest existing up-to-date one was arguably offered by Cumming in her book *The Sonic Self: Musical Subjectivity and Signification* (2000). Not presenting any sustained, thorough exegesis of interpreting musical works, Cumming's essay is chiefly an introduction and demonstration of semiotic tools applied to the field of musical performance, in particular Cumming's philosophical arguments and Peircean theories. The author presents her philosophy as, in her own words, "semiotic", in that sounds are not presented merely as acoustic phenomena, but as something capable of carrying connotations, including those human "subjective" qualities stemming from voice, gestures and actions. Besides, Cumming's insights are importantly influenced by Peirce's interest in the play of signs, and by the whole area of feminist music theory (Marion Guck and Marianne Kielian-Gilbert, a.o.). Perhaps in relation to this, a significant part of Cumming's thinking on the performer's identity is devoted to her bodily signs.

The use of Peirce's theory of signs, in Cumming's essay, is not necessarily literal, particularly the structure he gives to the sign itself, which is here manipulated in order to find a language for describing an area of artistic activity whose creations are neither purely material nor purely mental. Cumming illustrates this structure by reconsidering the practical exercise of producing sound on the violin, then giving it a quality of "singing" in playing a "cantilena" style (it must be said that the author restricts herself to only "singing" or weight of tone/sound, which of course are not the only qualities within performance):

Bodily movement has to become a vehicle for creating sounds that call to mind something they do not "literally" possess, the qualities of a singing voice. In Peirce's terms, the material qualities of the sound are the sign vehicle, by which it comes to represent (to be a 'representamen' or 'sign'). The vocal grain it achieves is its 'object', what it stands for. Without interpretation, no material sound produced by a non-human instrument can be heard as a voice, hence, in the third logical position there is an 'interpretant'. It acknowledges two things: the conventions that allow a violin's sound to be heard as vocal in some contexts, and the act of recognition in a particular moment of listening. What transforms a dead, mechanical performance into a 'live' one is the creation of sound as a sign—in this case, a sign for 'singing' in an appropriate tone (Cumming 2000: 29).

In the present article, the intention is to demonstrate that Peirce's theory is useful in the field of the performer's corporeality, in that it helps us distinguish between the different ways a performer's physical gesture may refer to the actual "text" that is being performed.

(1) An icon, which is a relation of physical resemblance, may appear whenever the performer *translates* (often synaesthetically) a physical quality of the opus into a bodily gesture. Raising up with one's body (or perhaps with the neck only) when high notes are being played; parodying an angry expression in correspondence of *forte* dynamics and/or March tempos; getting closer to the keyboard as if to hear better when the dynamics is *pianissimo*… all these are gestures of an iconic type, in that they are forms of "portrayal" (and adaptation to ordinary, non-musical life) of the musical events.
(2) An index, a relation of physical contiguity, is by far the relation mostly rooted in the biology of any semiosis, because the representamen is simply the natural "consequence" of its object, such as in the ever-employed example of smoke and fire. In the case of musical performance art, the music does not only *suggest/inspire* gestures to the performer, but it can actually *drive/force* him/her to move in a certain manner. A *fortissimo* dynamics, for instance, requires a certain muscular tension that produces a particularly rigid, as if *nervous*, posture; passages that require an inverted position of right and left hands require certain torsion in the body, and so forth. Obviously (and it happens rather often), the performer may decide to emphasize such gestures (e.g., a longer-than-necessary run-up of the hands towards the keyboard before the final cadenza), for the sake of a bit of extra-spectacle, or perhaps to feel him/herself more engaged in the performance, but in general this kind of relation represents the most natural one in Peirce's whole categorization. For this reason, and for the purposes of the present classification, we may here welcome a proposal that was formulated in the field of zoosemiotics:

> Despite the fact that, in the Peircean system, the icon is the sign-object relation connected to firstness (i.e., the mode of being of what is without reference to any subject or object), the analysis [should] start from indexes […]. The […] logocentric nature [in human

communication] predictably manifests a greater interest in symbols and icons. In a sense, one may go as far as to say that one of the main side-effects of the production of an artificial and/or cultural sign system is exactly its emancipation from indexicality, in favour of iconicity and (most of all) symbolicness. (Martinelli 2010: 69–70)

Although coming from a completely different field, this statement holds true for musical performance as well, in that indexical gestures are the only ones that are not mediated by any cultural/conventional choice of the performer (save the mentioned spectacularizations, which are anyway mere forms of emphasis, not deception, as compared to the real effort), but simply stem from his/her biology and anatomy, in relation to the physical demands of a musical piece.

(3) Finally, symbols, relations of conventional/cultural type, appear every time a performer displays a gesture that has been codified by some sort of social agreement, or rule, and that anyway bears no apparent similarity with the object. As pointed out above while talking about behavioural codes: performers (at least, most of them) may, for instance, keep a rather erect position while sitting at the piano, because that is the requirement of the etiquette of décor and elegance in a concert hall. Hands might be left suspended in the air for a few seconds at the end of a piece to convey the long-codified message that the audience should not applaud yet, etc.

9.3.2 Performance Gestures According to Roman Jakobson's Model of Language Functions

In what follows, Peirce's model is combined with Roman Jakobson's model on language functions. The goal is allowing sign relations and sign functions to interact in my classification of performance gestures. On the one hand, the interest here is an understanding of *how* a given gesture refers to the music performed (which is what Peirce provides); on the other, it is equally important to have an idea of *what* these gestures are *for*. In this case, the gesture may address the musical work itself, but not necessarily: it could be the performer, the audience or another type of context, and this context might even not be present there and then. The result of this combination shall be summarized and explained in a specific table (see Table 9.1). First, however, some information on Jakobson's model is pertinent.

As widely known, Jakobson (1960: 353–357) identifies six main functions within a communication system, and each of them is in turn related with a particular element of semiosis: the expressive function refers to the sender, the conative refers to the receiver, the phatic refers to the establishment of contact between sender and receiver, the referential refers to the context of the message, the metalinguistic refers to the code employed in a given communicative instance, and the poetic refers to the formal structure of the message. These functions are far from being isolated from each other or mutually exclusive: in normal conditions, a message is

Table 9.1 Possible sign typologies of a performer's gesturality

Icon	Index	Symbol
Translating a physical quality of the work into gesture	The work drives/forces the performer to move in a certain manner	Displaying an unrelated but usually conventional gesture
Expressive—Displaying the emotional state of the performer		
Conative—Addressing/affecting the audience, also in a "institutionally-codified" way		
Phatic—Establishing an intellectual and emotional bond with the audience		
Referential—Relating to context (instrument, hall, orchestra members…)		
Metalinguistic—Addressing (rather than following) movements and codes		
Poetic—Embellishing or emphasizing the normal movements required for playing		

the intersection between two or more functions, even if one of them is often dominant and more evident. In detail, and in relation to the art of musical performance, these could be exemplified as follows:

(1) The expressive function concerns that use of communication in which the most relevant part is the display of the emotional state and the identity of the sender. Certainly most, if not all, messages imply an expressive component, but clearly emotional elements can be more evident in certain cases than in others. The performer who closes the eyes and gently rocks the head to a particularly melodic and dreamy passage, is undoubtedly putting a particular accent on the emotional dimension of his/her corporeality.

(2) The conative function concerns that use of communication in which the sender uses the message in order to make the receiver have a certain reaction and act consequently. The receiver, in our case, is anybody who is exposed to the performance, for instance, the audience of a concert hall or (although the interaction in this case is indirect and artificially mediated) the listeners to a recording. The degree of communication between performers and audience (which shall be explored more specifically when talking about phatic signs) may vary depending mostly on the performer, and the history of piano playing has offered a whole catalogue of types, ranging from the nearly autistic performances of Grigory Sokolov or Radu Lupu (with almost no interaction with the audience), to the histrionic quality of Lang Lang's concerts (not to mention other musical idioms, such as rock, where the practical absence of a codified etiquette of behaviour and restrictions allows performers all sorts of interaction with the audience, including the infamous practice of stage-diving). And, on the other hand, the concept of spectatorship, the behaviour of concert attendants and their interaction with performers have been subject of investigation for several scholars, among whom, besides Small (1998) and Pitts (2005) should at least be mentioned. At any rate, and going back to performers, there are bodily signs and gestures that are "institutionally" conative, meaning that they are part of a codified set of rules in which the performer solicits a certain reaction (or lack thereof) from the audience. One already mentioned example is the

"suspended hands" signal that requires additional seconds of silence after the piece is over. Additional gestures include a prolonged time of preparation, on the part of the performer, *before* the performance, if the audience is not yet completely attentive and silent (prolongation normally manifested by removing the hands from the keyboard, to clearly show that the performer will not start playing unless everybody keeps quiet); the occasional interruption of a performance if the audience's "coughing level", so to speak, exceeds an acceptable threshold (interruption that, in a couple of cases, may be implemented with a severe spoken reproach to the audience[16]); and others.

(3) The phatic function refers to the use of communication mainly aimed to establish and/or keep contact between sender(s) and receiver(s). That goes from the simple act of greeting each other, to the most complex activities aimed at cementing an interpersonal or group relationship. To this extent, not surprisingly, it is in the nature of performance itself that of creating a phatic communication: after all a performance is exactly the act of displaying the ideas and the efforts of a composer (not yet conveyed to anybody except him/herself) for the consumption of other subjects. That is: the performer is the very figure who establishes *a contact* between musical work and listeners, who allows an actual communication between the two parties. Less generally, we can also think of the various gestures that the performer does in order to activate and/or reinforce the bond with the audience, starting from the fact itself of bowing in front of it, before and after the performance, up to those musicians who are particularly eager to have eye-contact with the audience, or to add additional non-verbal language to entertain and/or please them. Of course, a distinction should be made between gestures that are explicitly aimed at the audience, and gestures that are simply enacted for the performer's own sake (concentration, emotional engagement, or something else): however, to an extent, the distinction is irrelevant, because, in order for a phatic function to be defined as such (or any other of Jakobson's functions, for the matter), there is no need for the sign to be *intentionally* aimed at establishing/reinforcing the contact between sender and receiver.[17] The unusual catalogue of movements that a pianist such as Olli

[16]Famous are the difficult character and maniacal demands of the likes of Keith Jarrett, who, on more than one occasion, has not hesitated to interrupt his performances if the least, imperceptible sound was produced by the audience. There is no discussion, at this point of Jarrett's career and fame, that this attitude has contributed to his public personality, in all kinds of directions (from veneration to irritation, up to the audience's comprehensible terror to even produce the occasional, and traditional, cough). Among the classical pianists, Ivo Pogorelich is known for unpredictable reactions to any possible distractions on the part of his audiences; a rather harsh 'morale' by another pianist, András Schiff, to a coughing audience of his all-Beethoven recital at Helsinki Festival in the year 2007 was personally witnessed by the author.

[17]That of intentionality in semiosis (communication, in particular) is by far one of the most critical questions, and a great matter of contention among semioticians, particularly those interested in the establishment of the so-called "semiotic threshold". I do not intend to participate in the debate, here, but simply point out that, intentionally or not, performers and audiences are connected by several instances of signification.

Mustonen displays during his concerts may certainly be the result of his own needs (some would say neuroses), but that does not exclude the impact that these movements have on the crowd, from surprise to amusement, up to sheer annoyance. In addition, it must be also said that the amount and quality of contact between performers and audiences is also heavily affected by the context. On the one hand, as mentioned already, a difference in musical context may mean a great deal, and there is no need to repeat that a rock pianist, as compared to a classical one, can take as many liberties as s/he wishes, when it comes to interacting with his/her fans. But even remaining within art music, there is a great difference between performing in Carnegie Hall or in, say, educational concerts for children, where a lot of comedy routine is not only allowed, but in fact encouraged. Not to mention contemporary and experimental music, which have often used interaction with the audience as one prominent artistic feature. Finally, although a decision was made to focus the examples in this and the previous category (conative signs) on the performer-audience interaction, it is perhaps useful to remark that this is not the only possible communicational bond. For instance, in contexts where the pianist is performing with other musicians, there is also a great deal of contact with the colleagues, starting from the fact itself of synchronizing the beginning of a performance (normally executed with facial expressions and eye-contact), up to the final hand-shaking moments.

(4) The referential function describes that kind of situation where signs refer to a context, whatever that context may be (surrounding environment, subjects, time units, etc.). Such sentences as "Johann Sebastian Bach was born in 1685", or "The Metropolitan Opera's auditorium is one of the largest in the world, seating over 4000 people" are all of referential type, as is also any musical work's title with descriptive features (*The Four Seasons*, *Prélude à l'après-midi d'un faune*, etc.). Obviously, music itself is much more referential than some old school musicologist would like to think (including entire genres or compositional techniques that were explicitly born with a referential intention, such as the canon or the saltarello). In the specific case of performance, referentiality may appear in many different gestures, aimed at different contexts: the work itself, the concert hall, other musicians, etc. To an extent, every musical work offers the performer an opportunity to describe it through his/her corporeality: the body often goes along with temporal units, melodic passages, technically challenging moments, etc. Some works are of course more prototypical than others, in that their musical properties are already descriptive in se, so it is more natural for the performer to accord with them rather than more abstract pieces. Evidently, something such as Camille Saint-Saëns' *Carnival of Animals* offers more referential opportunities than many others.

(5) The fifth type of communication function is called metalinguistic, and it is probably the most complex to describe in relation to a performer's corporeality. It occurs when signs are used to refer to other signs, rather than straight away to objects, creating an additional layer of semiosis, and therefore requiring additional interpretive effort. Metalinguistic communication typically occurs in

instances of playing, deception, art and ritualization. A metalinguistic quality is for instance instrumental in order to create those elements of ambiguity that Umberto Eco considers a crucial ingredient in artistic communication (the famous notion of "Open work", cf. Eco 1989). That applies to music as well, but, of course, corporeality in musical performance is not such an obvious application of the concept. Yet, examples can be found. Glenn Gould's already mentioned habit of *conducting* his performances with one arm is a good case, because it is a form of gesticulation that is not proper to a pianist, but to another key-figure of musical performance, namely the conductor: a sign, thus, that refers to another sign, before addressing a specific object (i.e. regulating/pacing the musical flow). Also, whenever a gesture/act *addresses* any form of code, or even ritual, within the environment of performance, rather than simply *following* that code, chances for metalinguistic semiosis are high. Such convention-challenging performers as Ivo Pogorelich may have made, in the famous case of the Chopin competition, an anti-traditionalist statement by dressing in a very informal, inelegant, way. The sign (dressing informally) in such a case refers to another sign (the dressing code) to convey a message that in fact concerns the pianist's skills as a performer, not certainly his opinions on fashion (that is, challenging the traditional dressing code because of challenging also the traditional way of playing).

(6) Finally, a poetic function in communication focuses on the message itself, elaborating on its representation and form, rather than its contents. This is probably the most conventional bodily sign detectable in performers, along with expressive ones. Poetic are all those gestures that embellish or emphasize the normal movements required for playing. A hand approaching the keyboard, or leaving a certain area of it to reach for another, or lifted up before descending again on the keys… all these are gestures that a performer may execute in a rather straight, economic, manner. Yet, many pianists like to embellish those movements by making them more harmonic, perhaps a bit slower, sometimes even tracing imaginary doodles in the air. The same applies for any other movement of any other body part (head and facial expressions being very typical cases).

9.4 Concluding Remarks

Having explained in general terms how these particular theories by Peirce and Jakobson work within the field of performers' corporeality, the intention, as anticipated, is to suggest a combination of the two, for the purposes of encompassing the topic into a single model. The reasons for this decision lie in the fact that the two theories are perfectly compatible, both methodologically and, as far as my research goes, empirically. Indeed, on the one hand, Jakobson, through his classification, answers to the most obvious, yet unavoidable, question: what are these gestures for? What kind of role they have in a performer's ethology? On the other hand, Peirce aims to illustrate how signs refer to their objects, that is, what kind of configuration their relation takes. We need this, because we are trying to analyze the intrinsic semiotic nature of these gestures, and the relation sign-object is the most important one in any process of semiosis. It is my claim that by combining the two theories one shall be able to encapsulate different forms of analysis of performers' corporeality, including the existing ones that were mentioned in the literature review above.

In Table 9.1, Peirce's triad is placed on the top row, and Jakobson's functions go on the first column, so that their combination produces eighteen possible sign typologies. To take just a couple of examples, the above-mentioned act of emphasizing the natural gestures of performance is a typology that, under usual circumstances, occupies the intersection between the poetic function and the indexical relation; a pianist lifting his neck or looking up with his/her eyes in correspondence of high notes is, in normal conditions, a combination of the expressive function and iconic relation; and so forth. It is to be emphasized "in normal conditions" because with categorizations of this type there is always the risk of sounding too schematic and ignoring the several grey areas that exist across categories. This aspect will not be overlooked, but it is certainly hoped that some direction to an indistinct aggregation of apparently ambiguous or insignificant signs has been provided.

References

Barthes, Roland. 1977 [1972]. The grain of the voice. In *Image-music-text*. Essays selected and translated by Stephen Heath. 179–189. New York: Hill and Wang.

Butler, Judith. 1990. *Gender trouble: Feminism and the subversion of identity*. London: Routledge.

Clarke, Eric F. 1998. The semiotics of expression in musical performance. *Contemporary Music Review* 17 (2): 87–102.

Clarke, Eric F. 2004. Empirical methods in the study of performance. In *Empirical musicology: Aims, methods, prospects*, ed. Eric Clarke and Nicholas Cook, 77–102. Oxford: Oxford University Press.

Clarke, Eric, and Nicholas Cook (eds.). 2004. *Empirical musicology: Aims, methods, prospects*. Oxford: Oxford University Press.

Cumming, Naomi. 2000. *The sonic self: Musical subjectivity and signification*. Bloomington, Indianapolis: Indiana University Press.

Davidson, Jane W. 1993. Visual perception and performance manner in the movements of solo musicians. *Psychology of Music* 21: 103–113.

Davidson, Jane W. 1994. What type of information in conveyed in the body movements of solo musician performers? *Journal of Human Movement Studies* 6: 279–301.

Delalande, François. 1988. La gestique de Gould: éléments pour une sémiologie du geste musical. In *Glenn gould, pluriel*, ed. Ghyslaine Guertin, 83–111. Verdun, Quebec: Louise Courteu Editrice.

Eco, Umberto. 1989. *The open work*. Translated by Anna Cancogni, with an Introduction by David Robey. Cambridge, MA: Harvard University Press.

Elsdon, Peter. 2006. Listening in the gaze: The body in Keith Jarrett's solo piano improvisations. In *Music and gesture*, ed. Anthony Gritten and Elaine King, 192–207. Aldershot: Ashgate.

Fabian, Dorottya, Renee Timmers, and Emery Schubert (eds.). 2014. *Expressiveness in music performance: Empirical approaches across styles and cultures*. Oxford: Oxford University Press.

Gaser, Christian, and Gottfried Schlaug. 2003. Brain structures differ between musicians and non-musicians. *The Journal of Neuroscience* 23 (27): 9240–9245.

Hatten, Robert. 2004. *Interpreting musical gestures, topics and tropes: Mozart, Beethoven, Schubert*. Bloomington: Indiana University Press.

Jakobson, Roman. 1960. Linguistics and poetics. In *Style in language*, ed. Thomas Sebeok, 350–377. Cambridge, MA: M.I.T. Press.

Jankélévitch, Vladimir. 2003. *Music and the Ineffable*. Translated by Carolyn Abbate. Princeton, Oxford: Princeton University Press.

Koeneke, Susan, Kai Lutz, Torsten Wüstenberg, and Lutz Jäncke. 2004. Long-term training affects cerebellar processing in skilled keyboard players. *NeuroReport* 15 (8): 1279–1282.

Kopelson, Kevin. 1996. *Beethoven's kiss: Pianism, perversion, and the mastery of desire*. Stanford: Stanford University Press.

Krings, Timo; Rudolf Töpper, Henrik Foltys, Stephan Erberich, Roland Sparing, Klaus Willmes, and Armin Thron. 2000. Cortical activation patterns during complex motor tasks in piano players and control subjects. A functional magnetic resonance imaging study. *Neuroscience Letters* 278 (3): 189–193.

Livingstone, Steve R., William Forde Thompson, and Frank A. Russo. 2009. Facial expressions and emotional singing: A study of perception and production with motion capture and electromyography. *Music Perception* 26 (5): 475–488.

Martinelli, Dario. 2010. *A critical companion to zoosemiotics*. Berlin, New York: Springer.

McCaleb, J.Murphy. 2014. *Embodied knowledge in ensemble performance*. Farnham: Ashgate.

McClary, Susan. 1991. *Feminine endings: Music, gender and sexuality*. Minneapolis: University of Minnesota Press.

Merriam, Alan. 1964. *The anthropology of music*. Evanston, Illinois: Northwestern University Press.

Mitchell, Mark. 2000. *Virtuosi—A defense and a (sometimes erotic) celebration of great pianists*. Bloomington: Indiana University Press.

Nattiez, Jean-Jacques. 1990. *Music and discourse: Toward a semiology of music*. Translated by Carolyn Abbate. Princeton, New Jersey: Princeton University Press.

Navickaitė-Martinelli, Lina. 2010. *Pokalbių siuita. 32 interviu ir interliudijos apie muzikos atlikimo meną* [*Suite of Conversations. 32 Interviews and Interludes on Music Performance Art*]. Vilnius: Versus aureus.

Navickaitė-Martinelli, Lina. 2014. *Piano performance in a semiotic key: Society, musical canon and novel discourses*. Helsinki: Semiotic Society of Finland.

Pitts, Stephanie. 2005. *Valuing musical participation*. Farnham: Ashgate.

Schutz, Michael. 2008. Seeing music? What musicians need to know about vision. *Empirical Musicology Review* 3 (3): 83–108.

Small, Christopher. 1998. *Musicking: The meanings of performing and listening*. Hanover, London: Wesleyan University Press.

Tarasti, Eero. 2000. *Existential semiotics*. Bloomington, Indianapolis: Indiana University Press.

Tarasti, Eero. 2005. Existential and transcendental analysis of music. *Studi musicali* 2: 223–266.

Tarasti, Eero. 2012. Existential semiotics and cultural psychology. In *The Oxford handbook of culture and psychology*, ed. Jaan Valsiner, 316–341. Oxford, New York: Oxford University Press.

Thompson, William Forde, Phil Graham, and Frank A. Russo. 2005. Seeing music performance: Visual influences on perception and experience. *Semiotica* 156: 203–227.

Thompson, William Forde, and Frank A. Russo. 2007. Facing the music. *Psychological Science* 18: 756–757.

Virtanen, Marjaana. 2007. *Musical works in the making. Verbal and gestural negotiation in rehearsals and performances of Einojuhani Rautavaara's piano concerti.* Turku: Turun Yliopisto.
Wanderley, Marcelo M. 1999. Non-obvious performer gestures in instrumental music. In *Gesture-based communication in human-computer interaction: International gesture workshop, GW'99 Gif-sur-Yvette, France, March 17–19, 1999 proceedings*, eds. Annelies Braffort et al., 37–48. Berlin, Heidelberg: Springer.
Williamon, Aaron (ed.). 2004. *Musical excellence: Strategies and techniques to enhance performance*. Oxford: Oxford University Press.

Videos

Kempff, Wilhelm. 2003. *Schumann, Beethoven*. EMI Classics DVD, B0001AW052.
Mercury, Freddie, and Montserrat Caballe. 1988. *La Nit festival at Montjurich Castle in Barcelona (8/10/1988)*. https://www.youtube.com/watch?v=OPizyP4x30I/ Accessed 2 August 2017.
Quimby, Fred (prod.), William Hanna and Joseph Barbera (dir.). 1947. Tom and Jerry, Episode 29. *The Cat Concerto*. https://www.youtube.com/watch?v=uKZgi06fVsk/ Accessed 2 Aug 2017.

Chapter 10
Food Communication and the Metalevels of Carnism

Dario Martinelli

Abstract The present essay offers some reflections concerning the cultural and ideological debate between veganism and carnism. The increasing popularity of the vegan lifestyle among diverse layers of society, and most of all its transition from a hardly-tangible phenomenon to a very visible and outspoken one, has generated a strong reaction from those who Melanie Joy has defined "carnists" (i.e., meat-eaters not only in the culinary sense, but also in the cultural one), but it has also brought to the fore the question of the different communication strategies of the two parties. After a brief but hopefully-exhaustive introduction to veganism as socio-cultural and ideological movement, the essay proceeds to focus on the communication strategies operated by carnism to convey or disguise (depending on the case) the kind of information that veganism is ethically pointing the finger at: the promotion of meat as product, necessity, display of luxury, nutritionally-valuable aliment, and so forth. In describing such cases, and offering a theoretical framework for their interpretation, it is also argued that carnism, like any other form of anthropization, is filtered via three similar-but-compatible mental attitudes: anthropocentrism, speciesism and anthropocracy.

10.1 Introduction

The social and political relevance of the debate veganism-carnism is nowadays so plain to see that it probably needs no specific argumentation. As discussed extensively in Berkmanienė-Martinelli 2018, the statistics related to the exponential increase of veganism in the last few years display a number of features whose significance have much more to do with ideological and cultural aspects than with culinary or nutritional ones. To mention just a few, vegans tend to be found in a significantly higher number among women, liberal-leftists, atheists, urban citizens and highly-educated people (Berkmanienė-Martinelli 2018). While there may be

D. Martinelli (✉)
Kaunas University of Technology, Kaunas, Lithuania
e-mail: dario.martinelli@ktu.lt

A. Olteanu et al. (eds.), *Meanings & Co.*, Numanities - Arts and Humanities in Progress 6, https://doi.org/10.1007/978-3-319-91986-7_10

different (and sometimes contradictory) explanations for these data, it is undeniable that they represent an ideological/political/cultural body of information, and that the "staying light and healthy" preoccupations are hardly the only (or often the main) drive behind the decision to embrace the vegan lifestyle.

By consequence, and predictably, *not to* embrace veganism has an become ideological/political/cultural affair. Until very recently, there had been two traditional terms and concepts that the word "vegan" was usually designated to oppose. One is "omnivore" and the other is "carnivore": both were (are) wrong, scientifically and semantically. "Omnivore" is a term that defines those animals, for instance, the human ones, whose diet can potentially include both vegetable and animal aliments (and of course some fungi as well—we are using here the gastronomic principle that classifies mushrooms as vegetables). Human beings, in the course of their evolution, have experimented different diets, and, from a certain point onwards (ca. 2.6 million years ago), have adapted to eat other animals as well. It was an important turning point, also in terms of social organization and development of intelligence, which made them flexible from a nutritional point of view, and transformed them accordingly from a physiological and anatomical point of view: to make one example, the human jaw has neither a carnivore nor a herbivore structure, but it is somehow a compromise between the two. Some aliments, in the human diet, are more essential than others (for instance, chances to survive are dramatically higher if one eats only vegetable food, than only animal food), and it is not to be excluded (in fact, it is very likely), that evolution in the future will lead our species "somewhere else", adding food we are currently not accustomed/adapted to, or subtracting/replacing others that seem to be indispensable now. There are, however, two problems. The first is that "omnivore" is a word deriving from two Latin terms that stand for "eat" and "all". Human beings are animals who eat only some vegetables and some animals (and some fungi too), and yet they are described as *omni*vores. Several vegetables and animals (not to mention minerals, metals, etc.) are not digested by humans, or they are even lethal. There is even a word, "pica", that defines the set of psychological disorders of eating things we cannot eat (that is, paradoxically: a "omnivore" can be sick of wanting to be a *real* omnivore, that is, "eat everything"): hyalophagia (appetite for glass), lithophagia (appetite for stones), plumbophagia (appetite for lead), and so forth (up to the more gothic-literary hematophagia, or vampirism), are all life-threatening disorders that prove that our species is very far from qualifying for a prefix like "omni-", as applied to its eating capacities.

The other problem is that a word like "omnivore", at least scientifically, is much more centered on adaptation than needs. "Omnivore" is an animal who *adapts* to eat a great variety of aliments, not one who *must* eat them all. Exactly due to their flexibility, human beings can manage to survive in many different dietary conditions, including those that *exclude* a significant amount of the food they can eat. Nothing new under the sun, here: as a matter of fact, millions and millions of (or perhaps all) the inhabitants of this planet survive without eating some or many of the aliments that they could eat. Some for allergy or intolerance; some for ideological, cultural or religious reasons; some for the objective impossibility to find certain food in certain places (including, of course, financial impossibility). Arguably, no human being eats *all* s/he could eat. Many also do not eat things they *should* eat (because of their crucial nutritional value), and nevertheless many of these manage to survive anyway.

What happens, thus, is that both these characteristics (flexibility and adaptation) end up being "quasi-secret" to our cultural perception, thanks to a word, "omnivore", whose inclusive semantic implications (as opposed to "vegetarian" and "vegan", which are exclusive) appears as describing an individual who can and *must* eat everything. However, while there is a social acceptance of allergies/intolerances (as medical conditions) and religious choices (as culturally undisputable), the subject who chooses to exclude certain food from his/her diet for ideological/ethical reasons, is often confronted with the "value" of omnivoreness. "One must eat a bit of everything" is one of the recurrent refrains that vegans/vegetarians are exposed to, on a fairly daily basis. If you *can*, you *must*. If you must, you *want*. If you do not want, you are wrong.

So: do we have terminological, fairer, alternatives to "omnivore"? Of course we do, and not by chance, within the scientific communities, those alternatives are more and more utilized. These are "eurifagous" and "polyfagous", two words, deriving from Ancient Greek, who put an accent on the "variety" and the "quantity" of food we can intake, but do not pose the question in absolutistic terms. As eurifagous and/or polyfagous, we are defined as animals who can eat "diverse" and/or "many" aliments, not all of them. If scientific accuracy is what we are after, those terms should be preferred, without hesitation, to "omnivore".

Now. The second problematic word is "carnivore", this one usually adopted in demagogical or a (self-)ironic manner. Vegans and vegetarians often use the word "carnivore" to qualify polyfagous subjects who eat meat, with tones that range from irony to contempt. Meat eaters themselves may sometimes use this word, in application to their own eating habits, in an attempt to emphasise (occasionally with humor) their nature as individuals belonging to a species that has achieved significant progress in its evolution, by learning to hunt and breed, and by supposedly reaching the dominant position in the food chain.

The word is obviously fallacious, in its application to the human species. Humans are not carnivores, as well as they are not herbivores, due to a consistent amount of physiological, anatomical, ethological and psychological characteristics that one should refuse to list in an academic article, since they are so much "popular knowledge".

Something, however, in the employment of this word (and here we return to the initial point), suggests that the assignment of semantic categories, when it comes to eating habits, is not dissimilar to the way we can define "gender" as something determined by both natural and cultural processes. That is: in a world where giving up meat can be, and often is, the result of an ethical-ideological standpoint, it follows that also "not giving up" can be, and often is, ideological. And that is particularly true in an epoch, the last ten years, when the confrontation between the two lifestyles has become visible, open and not rarely tough, and when the two parties are not so unbalanced in number as they used to be: while meat-eaters remain the majority, indeed, the counterpart is no longer composed by few isolated and disorganized weirdoes: vegans are now a tangible, vigorous and educated community gathered in associations, forums and social network groups, who are also endowed with the "bonus points" of vocal VIPs and influencers, increasingly-informed physicians, scientists and intellectuals, more institutional acknowledgment, and world-wide spread internet videos disclosing the conditions of farm animals or the environmental impact of the meat industry.

Social psychologist Melanie Joy has devised an instantly-successful term to define meat-eating in its ideological, social and cultural sense: carnism (Joy 2001). Its primary aim is the description of the psychological fact that the perception of animal products as food (or as anything else, really) is highly cultural. Different cultures in the world may apply different categories of perception to the same species: a dog may be a beloved pet in some societies and delicious food in others. A cow may be exploited in all possible ways in some countries, and be a sacred animal in others—and so forth. Specifically in relation to meat, and despite the existence of millions of animal species on the planet, it is also interesting that, all in all, human beings eat on a regular basis only few (mainly herbivorous) of them. In that sense, like all ideologies, carnism too must be understood as a "shared set of beliefs, as well as the practices that reflect these beliefs" (Joy 2010: 29).

Since my competences, as author of this essay, are mostly rooted in the areas of semiotics, animal ethics and cultural studies, I am interested in focusing on the modes and manifestations of carnism as communicative strategy—be that marketing or other forms of persuasive communication.

10.2 The "Filters" of Carnism: Anthropocentrism, Speciesism and Anthropocracy

If carnism is, so to speak, "ideological material" (and not only do I agree with Joy that it is—I also think that all other anthropization[1] processes are), then there must be also some (at least) ontological and existential factors that de facto mediate it.

[1]I use here the wider definition of "anthropization" that is discussed in Celentano-Martinelli (2018). Since that essay is forthcoming, at the time I am writing the present one, I shall summarise the main points here. The study of free/wild non-human animals is there identified as a study of "An-anthropization" (the condition of total absence of anthropization) or—when some form of human impact/influence/modification is likely or bound to happen—"Pre-anthropization" (the temporal condition antecedent to anthropization). As soon as an actual anthropizing process occurs, we can classify at least 22 different types, each of them generating a number of processes and practices, one of which is indeed carnism. They are: (1) Ab-anthropization (A. developed apart from humanity); (2) Anthro-anthropization (A. aimed at anthropomorphizing—physiologically, ethologically, culturally, etc.—the non human animal); (3) Anti-anthropization (A. developed autonomously by non human animals, which actually damages humanity in more or less serious ways); (4) Archeo-anthropization (A. developed in pre-historic times, often as results of co-evolution); (5) Auto-anthropization (the non human animal, so to speak, "volunteers" to be part

I shall not focus here on the needs (or sometimes just wishes) that motivate and activate carnism (these, also, to be considered mediating factors), but rather on the role humans ascribe to themselves as agents of anthropization in carnistic processes. This role is, in my opinion, characterized by three main domains—often overlapping in action, but not in meaning: anthropocentrism, anthropocracy and speciesism. According to the usual definitions (e.g., Bartolommei 1995: 40–83, or the semiotic re-elaboration of the latter in Martinelli 2010: 302–304), anthropocentrism interprets Nature as (a) an entity existing *apart from* and *for the benefit of* humans, so that (b) nothing in Nature can be considered in itself, autonomously from humans; and (c) it is ethically acceptable for humans and non-humans to be treated in different ways. In other words, Nature is not of interest because of its hypothetically *intrinsic value*, but just because of its *instrumental value*, i.e., the values it has for and to humans.

Speciesism is a term that was first coined by Richard Ryder in 1970, and then brought to general attention by Peter Singer in his classic *Animal Liberation* (Singer 1975—see also Martinelli 2016: 149–151 for an adaptation of the concept within the platform of Numanities). It is defined as the discrimination against certain animal species by human beings, based on an assumption of humankind's superiority. Its first appearance as a dictionary entry in 1985 (in the *Oxford English Dictionary*) marked the official acceptance of this word into common language, and—since then—its increasing popularity among activists and intellectuals engaged in animal advocacy (much to the detriment of the word "veganism" itself, which ended up designating more and more the culinary sphere only, as opposed to

---

of the human environment, accepting its dynamics); (6) Bene-anthropization (the human being is "favourable" to an A. process, even when it features elements of risk/damage); (7) Corpo-anthropization (A. that requires a significant manipulation of the subjects' bodily constitution, physiology, etc.); (8) Credo-anthropization (illusory form of A., that may reveal itself as fallacious); (9) Grapho-anthropization (written/visual A.); (10) Legi-anthropization (A. that occurs or changes status by means of juridical or scientific regulations); (11) Ideo-anthropization (A. occurring at ideological, cultural, mythical level); (12) Idio-anthropization (A. that occurs in a confrontational manner: the non human subject/s is anthropized out of fear or specific wish to subdue); (13) Liber-anthropization (A. within which the non human animal is allowed to follow his/her natural biology); (14) Logo-anthropization (A. due to linguistic dynamics); (15) Loco-anthropization (A. that is characteristic of certain contextual/environmental conditions and that is not possible in others); (16) Macro-anthropization (A. as "large", possibly global, phenomenon); (17) Micro-anthropization (A. as circumscribed, very local, phenomenon); (18) Philo-anthropization (A. due to emotional attachment, affection, sexual attraction, etc.); (19) Psycho-anthropization (A. that did not occur in reality, but only as a mental attitude and in some individual or cultural processes); (20) Semi-anthropization (partial A. in which the non human subject/s retain elements of their natural condition); (21) Sub-anthropization (A. that was a consequence of another anthropization); (22) Sin-anthropization (A. involving the anthropization of different species/specimens at the same time).

Needless to say, each entry should not be considered isolated from the others, but in fact often in the position to intersect and overlap with, contain or be contained by, other entries. If we for instance think about anthropization as generator of (or creator of the conditions for) carnism, we can see that actually pretty much all the twenty-two types may have had a role, in one way or another, or in one moment of another of the process.

the full-round ethically-filled significance it had at the beginning).[2] Much like racism tends to be considered a natural, though not compulsory, consequence of ethnocentrism, speciesism is certainly a product of anthropocentrism, although—similarly, and at least in theory—it is possible to be anthropocentric without being speciesist (for instance, one may say that Buddhist philosophy is of that kind).

Anthropocracy, finally, is here proposed as a new word—or rather a new semantic option for an old and nearly-unused word (coming from the medical jargon). Easily drawing from ancient Greek ("human" + "power/domination"), the term is here intended to signify the positioning that humankind takes within the environment and in comparison with other animal species: the positioning of the ruler, the one in charge of everybody else's destiny, and the one allowed to colonise and manipulate any desired portion of the planet. Already at an early stage of human evolution, domination became the main, and soon after the only, form of human adaptation to the environment. Keeping up with the previous (and very recurrent, particularly among animal activists) comparison with human intergroup dynamics, whereas anthropocentrism reminds of ethnocentrism and speciesism

---

[2]As readers may or may not know, the first official appearance of the term "vegan" is dated 1944, when a Leicester member of the UK Vegetarian Society, Donald Watson, led a sub-group of fellows who renounced also eggs and dairy and coined for it the word "vegan", also setting up a quarterly newsletter called *The Vegan News* (eventually called simply *The Vegan*). From the pages of this paper the first official definition of "veganism" appeared, in 1951: "The doctrine that man should live with-out exploiting animals". Thus, the origins of the movement are unmistakably ethical, and the dietary aspect is simply one of the implications of a life "without exploiting animals". Animal advocates are often outraged by the fact that the discourse on veganism has an almost-exclusive emphasis on food, and at the same time the anti-vegan community criticizes expressions like "Vegan shoes" or "Vegan cosmetics", thinking that they are only a marketing strategy to ride the wave of the increasingly-fashionable *eating* lifestyle.

Few years later, in 1960, the American Vegan Society's founder Jay Dinshah defined veganism by mentioning the Sanskrit word Ahimsa, "non-harming", accentuating again the ethical nature of the movement. It is only with the third, and still important, institutional acknowledge-ment of the concept that the attention shifts to the dietary side of the issue. In 1962, the Oxford Illustrated Dictionary includes for the first time the entry "vegan", defining it as "a vegetarian who eats no butter, eggs, cheese or milk" (Smith 2013: 197). The first significant form of "metalevel of representation" (the central concept of this essay, which I am going to define further on) had thus appeared, activating a process of mutation of the vegan discourse from the status of set of ethical claims to one related to health and culinary choices. Such process gained strength during the late 1960s, due to the New Age and hippie counterculture and most of all, in the 1970s, with the appearance of a significant number of medical studies on the benefits of the vegan diet—or alternatively on the damaging consequences of the meat-based diet (see for instance the overviews offered in Maurer 2002; Iacobbo and Iacobbo 2006; Freston 2011). At that point, the ethical area was getting occupied by increasingly popular terms such as "animal rights", "animal advocacy", "environmentalism", indeed "anti-speciesism" and others, and that virtually deprived veganism of its original significance, relegating it to the more circumscribed domains of nutrition, gastronomy and health. Since its debut in mainstream culture (in the 2010s), the word "vegan" has partly reacquired its philosophical identity of any item not involving animal exploitation (or at least any *item* that would or might, in normal conditions: "vegan shoes", "vegan leather", "vegan cheese", etc.), but its connotative dimension of "diet" remains its strongest feature in social perception.

reminds to racism, anthropocracy may be designated to correspond to imperialism/colonialism.

As militant as this assertion may sound, especially in an academic context, it is very difficult to imagine, at the present state of human evolution, just any form of anthropization that is not mediated by one or more of these three factors—and carnism certainly makes no exception (also because it is arguably the quintessential form of anthropization—as I have hinted in a footnote, and certainly one of the forms with the biggest impact on Nature). Even the most benevolent types of anthropization depart anyway (and at least) from a self-positioning, from the human agent's part, "at the centre" of the anthrozoological space (e.g., the *responsibility* to preserve a given species), or as "magnanimous despot" (e.g., *ownership* of a pet).

10.2.1 Anthrozoological Biases as Intergroup Biases

The question is problematic *in se*, but it gets even more so, when we consider that anthropocentrism, speciesism and anthropocracy are likely to filter (and consequently distort) the anthrozoological relation with a number of significant biases that bear a striking resemblance with the ones described in Allport (1954) first and then, most of all, Tajfel (1981) in relation to (human) intergroup dynamics.

Following Allport (1954), categorizing people into groups by identifying some common attributes or characteristics is an adaptive necessity. It reduces the amount of information to be dealt with and thus reduces the complexity of the whole social world. Whether we are talking about, say, a Barcelona or a Real Madrid fan, people interested in football fit into a category depending on whom they cheer for. Some consequences go along with this simplification strategy. When a football fan identifies him/herself with, e.g., Barcelona (or, better, with that group of people playing in or supporting Barcelona), then s/he perceives him/herself and the group members as different from other groups. That person may work with a Real Madrid fan, have the same position, love the same movies, and live in the same town. But the Barcelona fan will view the Real Madrid fan as significantly different from him/herself. If another co-worker was also a Barcelona fan, but lived on the other side of town and hated those movies, s/he would still be seen as more similar to the Barcelona fan than to the Real Madrid fan. This co-worker is in practice a member of the same "ingroup" to which the Barcelona fan belongs.

To most moderate extents, such a categorization leads to the belief that one's own group is different from other groups in a kind of special way. Simply put, people place themselves into the ingroup and all other people into outgroup/s. Typical is the case of news distribution in newspapers and magazines. What is found there is not "Ethiopian news", "Australian news", "Colombian news" and so on. The distinction is rather between "domestic news" (ingroup) and "foreign news" (outgroup). In our case, this state of affairs corresponds to "anthropocentrism"—the idea that the human "group"/species is different from other animals in a special way than any species is different from each other: a concept that is often referred to as "animal plus".

In the worst case, however, categorizations produce intergroup biases, i.e., a belief or beliefs that one's own group is in fact "better" than *all other groups* (and this is where speciesism comes into the scene), or has a kind of "right" to dominate/step over other groups (and this is where anthropocracy comes into the scene). This leads to an increased attraction to ingroup members and a devaluation of outgroup members, and takes the name of *minimal intergroup paradigm*, a concept pioneered by Tajfel (1981). Ingroup members tend to behave in more pro-social ways and be more cooperative towards other ingroup members than they are to outgroup members. Keeping up with our example, football fans are obviously more pro-social and cooperative towards the ingroup members than those of the outgroup. Rarely do fights occur between two Barcelona fans, but it is common to see fights between Barcelona and Real Madrid fans. Of course, if the context shifts towards a conversation about another topic, like certain movies, the two Barcelona fans (which, in our example, did not share the same taste for films) may easily be in competition with each other, as parts of a different ingroup/outgroup dynamic (loves-certain movies versus hates-certain movies). Apart from being, at least in this historical moment,[3] the main socio-psychological characteristics of the human perception of other animals, intergroup biases are also the most effective "facilitators" for carnist communication.

The other important intergroup biases are the following:

1. *Outgroup homogeneity effect*: outgroup members are not only seen as being different/inferior from the ingroup, but also as being more interchangeable with each other. They are perceived as being more similar in their characteristics, and behavior than are ingroup members. As we shall see later, for instance, human beings have often a tendency to address "animals" as one single cauldron, without necessarily paying attention to species differentiation. Or, similarly, they may address whole species as they would address single human individuals.
2. *Polarized appraisals of outgroup members*: ingroup members tend to evaluate outgroup members as more extreme in a variety of positive and negative personal dimensions, thus creating greater social distance between ingroup and outgroup members. For example, *potentially* dangerous animals become *very* or *surely* dangerous. And while no statistics in this world may ever prove that there are more chances to be attacked on a street by a dog rather than by a fellow human being, still, while crossing paths with a human and a dog, there is an expectation that the dog may act more aggressively than the human.
3. *Ingroup favoritism effect*: simply put, the ingroup member instinctively tends to favor members of the ingroup over those of other groups. The "would you rather

[3]Signs of a progressive change of mentality are of course increasingly evident, in society, and that is exactly one of the effects of the spreading of movements like veganism. It is predictable that a more and more *inclusive* approach to other species will characterise the human societies of the future. Whether this is likely to be a short-, mid- or long-term process is difficult to establish, but the chances of its completion are much more likely than only ten years ago.

save an animal or a human?" refrain-dilemma is often raised in discussions about the value of human and non human life. In general, if s/he wants to avoid being labeled as a heartless fanatic, the interlocutor is expected to answer "a human, of course!".

4. *Differences in causal attributions across groups*: ingroup members make different attributions of cause and blame depending on whether the target individual is a member of the ingroup or not; ingroup members make more favorable attributions about members of ingroups than those of outgroups; positive behavior is attributed to internal causes rather than external ones, and so forth. E.g., a human act of compassion is treated as "noble", or—indeed—"humane", while another animal's compassion is often interpreted as "instinct" or anyway "not meant in the same (high) moral sense that (only) humans possess".
5. *Differences in information processing and retention across groups*: people process and retain information differently with respect to ingroup/outgroup members; ingroup members perform a more elaborate and careful processing about fellow ingroup members; have a more accurate memory for personal information regarding fellow ingroup members; remember more positive information about fellow ingroup members; have favorable expectations of fellow in-group members; and unfavorable expectations of outgroup members. It is undeniable, for instance, that the care taken to analyze a human response to a particular circumstance is much greater than the one devoted to a non human animal reacting to exactly the same circumstance.
6. *Illusory correlation*: Infrequent, distinctive behavior is associated with outgroups even though the behavior is constant across all groups. For instance, through words and expressions like "brutal", "beastly", "like an animal", "like a beast", as employed to describe human crimes, we assign to the condition of animality a number of behaviors/attitudes that are not only typical of *both* human beings and numerous other species (like rape), but also those that are *proto*typical (when not species-specific) of the *Homo sapiens* (like war).

As quintessential manifestation of anthropization, carnism is a solid carrier of a great number (perhaps all) of biases that affect our anthrozoological relationships. As have already mentioned, biases are results of "adaptive necessities", so—as much as we may criticise or defend the use of such a strong term like "biases"—it is important to remember that carnism, in a modern, highly-informed society, would not be able to exist without them. Once established that they need and/or want to eat other animals, in certain amounts, with certain modes of production, within a certain economic logic, and so forth (all of these variables changed in time, of course, in one direction or another), human beings had to develop a form of ideological "adaptation" to explain (but also promote, justify, argue …) the choice. Time and space, in addition, have qualified these strategies of adaptation in different ways: we call them "biases" here (in an academic article with a predictably mostly-academic readership) and now (in the 2010s, as we witness a "peak" of veganism as ideology). In other periods and places, they may have been called "natural order of things", "roles assigned by God", "survival of the fittest", "sad but necessary", and so forth. Not forgetting (and apologies for falling into a circular argument) that "intergroup biases" were elaborated by Tajfel to describe *human* relationships, and even in this condition they had other names in the past ("natural order of things", "roles assigned by God", "survival of the fittest", "sad but necessary" are all expressions that have been employed to explain human slavery, eugenetics, racism, sexism and basically all forms of human discrimination towards other humans).

10.3 The Metalevels of Carnism

Wanting to summarise in one single expression the fundamental implication of what discussed so far, it seems clear that the level 0 of any discussion on carnism is the human loss of awareness of its biological nature and ontological condition of "animal". In nearly all fields of human and anthrozoological interactions, we observe a continuous process of detachment of human beings from their animality. To make a trivial example, calling somebody an "animal" does not normally produce a delighted reaction, despite the fact that we do nothing but make the undeniable, scientific assertion, that each *Homo sapiens* is a member of that taxonomic domain known as the Animal Kingdom. Yet, when using that word, we usually do not mean it in its denotative sense, but rather in one of the (several) negative connotative ones (being uncivilized, violent, irrational, sexually abusive, etc.). Our perception of reality influences the words with which we describe it, but in fact the contrary occurs just as often: words influence perception. And words, as applied to non-human animals, are often confused and demagogical. When we say "animals", we think of every animal species except the human one. Our biological roots are forgotten, and in the process, we effectively create a cognitive gap, identifying ourselves as different (and hence, as we have seen, better/superior). This kind of process is not just limited to single words or single socio-cultural dynamics: what we established is a whole anthropocentric/speciesistic/anthropocratic discourse, which has the immediate effect of separating the world (i.e., the perception of reality) into centre and periphery—identity and otherness. To discuss the idea of animal means also to discuss the idea of otherness. Non human animals are perceived as *the others* par excellence, i.e., as we hinted, they are to humans what, in the past—and somewhere in the present—women were to men, blacks were to whites, native populations were to settlers, and so on. Humans basically think, do, have, are, etc., what animals do not, have not, are not.

The creation and establishment of such discourse and such socio-cultural dynamics produce what I here call a "Metalevel of representation"—the idea, in other words, that one is led to deal with a given portion of reality through means that belong to another portion of that reality, one that may be only indirectly related—or even, sometimes, not related at all. Animal advocacy theorists and activists have often employed the expression "cognitive dissociation" to refer, for instance, to the fact that—quoting Melanie Joy—we "love dogs", but "eat pigs" and "wear cows". In my humble opinion, however, this concept appears to be quite problematic and in itself anthropocentric: to conclude that there is a "cognitive dissociation" in distinguishing among species, and expressing a different range of emotions/attitudes towards them, means to assume that "not considering different species in the same way" is a fallacy, and therefore we should adopt an "outgroup homogeneity effect" in order to consider all non human animals as members of a single group. This, as we have seen, is the very essence of anthropocentrism. Although it may not be the noblest of attitudes, distinguishing or even discriminating across species has, to my mind, nothing to do with cognitive dissociations, but simply with taste, culture, degrees of empathy and other factors. Loving deeply one person, and at the same time disliking or ignoring another one so much that his/her death would not provoke any sadness, is something that happens to many people, and it goes nowhere near a "cognitive dissociation": if this can happen in relation to one single species (*Homo sapiens*), we certainly should not be surprised if it happens in relation to two or more different ones (and—moreover—none of which being one *we* belong to).

On the contrary, "metalevel of representation" seems to me a more suitable expression, as it does not point the finger towards some kind of mental dysfunction, but rather towards the decision (conscious or not) to process a given phenomenon with the cultural/ideological/discursive/etc. tools that are not directly applicable to it, but with others that are less pertinent. In other words, and back to Joy's catch-phrase, if we love a dog and eat a pig, it is because we represent dogs according to a given set of metalevels (e.g., companionship, playfulness, affection …) and pigs according to others (e.g., food, product, nutrition …).

Carnism provides just the perfect example, as it gets represented (in practices of information, marketing and the likes) according to numerous and diverse configurations. Borrowing from, and slightly reworking, a model employed for a rather different topic (political music), but with the common goal of investigating ideological forms of representation (Martinelli 2017: 27–29), I suggest to classify the rhetoric of carnism in four categories of communicative space:

1. **Hidden/disguised** (see Fig. 10.1): Carnist communication is coded in such a way that the connection between meat and its actual living source is not noticed and instead transposed into a metalevel that bears no resemblance with a "living" item. This may occur for various reasons, including practical ones (comfortable packaging, cleanness, etc.), but the most common, in a strictly ideological sense, is the active wish, from the sender's part (seller, producer, marketer, etc.) to prevent/discourage, in the receiver (buyer, consumer) any cognitive and emotional connection between meat and its "victims". The steak we buy in a supermarket does not even vaguely resemble the animal it used to be. In our daily and automatic process of price- weight- and quality-checking, there is simply no place for reflection that that very steak was—not long ago—someone that used to breathe and walk. Many toddlers and youngsters react traumatically as they become aware of what exactly a steak is, and many

Fig. 10.1 Two examples of "hidden" carnist communication: a supermarket shelf with packaged meat and a pile of tuna cans

vegetarians and vegans declare that their lifestyle choice was made immediately after making this discovery. The disguising process operated by the meat industry is successful when it creates detachment and indifference. Meat is a product, a food item, not a former life—this is where the "metalevel" is introduced: now the consumer will evaluate that object only with the tools pertinent to the idea of "product", failing to catch the obvious relation steak-sentient being. It is not a matter of encouraging ignorance in the customer; it is rather that any reflection on that relation is discouraged. Instances of hidden/disguised communication are most typical in the great distribution and range from the meat cuts packaged in polythene and plastic, up to processed meat/fish sold in tin cans (the latter being possibly the most successful process of disguise, as—not rarely—customers end up believing that products like tuna are not, in fact, animals). A metalevel of this kind is primarily "anthropocentric", as it mostly promote a perception of the NH animal as "instrumental value".

2. **Ambiguous/antithetic** (see Fig. 10.2): In this case, the communicative strategy does not aim at *hiding* the connection meat-sentient being, but rather to create rhetorical solutions (particularly hyperboles, antitheses and personifications) that switch the perception to completely ambiguous (also paradoxical) emotional levels. Typically, this is the case of the "happy-to-be-eaten" NH animals that we see represented in so many advertisements. Anthropomorphic, cartoon-like, cheerful chickens invite families (with kids) to a fried chicken fast food

Fig. 10.2 Three examples of "ambiguous" carnist communication: the "happy to be eaten" pig and chicken, and the virile "real men eat beef" bull for a steakhouse

restaurant; masculine, body-built bulls invite "real men" to the steakhouse; fat pigs dressed like chefs happily invite customers to eat pork ribs … It is a metalevel that bears a clear connection with "speciesism", and not by chance the "happy meat" case reminds of similar representations practiced in human racist contents, as the examples of "happy slaves" in Fig. 10.3 clearly show.

3. **Clear/informative** (see Fig. 10.4): The communicative strategy makes "meat" a visible/tangible item, in a way that is informative but not overstated. The receivers need to become "aware" of the connection meat-animal, because they need to make choices, evaluations, etc. (for instance, they may want to know if the sandwich they are about to buy in a snack-bar contains chicken, pork, beef or fish). The information in this case is emotionally-neutral, and for instance is manifested (as we, again, can see in Fig. 10.4) in form of stylized silhouettes of the animal whose meat needs to be represented (often accompanied by stereotypical colors: pink for the pig, brown for the cow, blue for the fish … and of course a green leaf will stand for the vegan option). In this instance, too, we face an "anthropocentric" metalevel, which once again shifts the attention from the living, sentient being to a "product" as such.
4. **Exposed/Ostentated** (see Fig. 10.5): In this case the connection meat-animal is not only visible, but rather assertive and esclamative—*cried out loud.*

Fig. 10.3 Three old racist advertisements with the "happy slave" Afro-American serving white (middle/upper class) masters

Fig. 10.4 Six relatively "neutral" icons used in food packaging

Fig. 10.5 Two forms of "ostentation" of meat/fish

A marketing-related strategy of ostentation is that of fish-sellers who exhibit living fishes in order to emphasise the quality and the freshness of their products. And, indeed, how can one find a fish fresher than a living one. The techniques for keeping the animals alive may range from aquariums packed with—say—lobsters to plastic boxes where the fishes agonise in the open air, occasionally relieved by a splash of water. The suspicion is that this marketing strategy is feasible only with animals that are morphologically and ethologically distant from human beings, like indeed fishes or insects. These animals are too small, too silent, too different; their suffering is not dramatic enough, compared to a human's, to elicit our empathy. The metalevel, here, is once more anthropocentric: the fish was still turned into a product, only a different type of it—a product whose freshness is the most valuable feature, and as such it requires evidence (the fish being kept alive). That such evidence happens to be also the proof that a sentient organism is being tortured is something by now irrelevant, because indeed the fish is at this stage a "product" and not a sentient organism.

There are other types of ostentation, also related to aesthetic requirements. One may consider those monumental meat dishes, served in certain haute-cuisine restaurants or folkloric feasts, that present the animal in more or less their original form, including head and legs, with a series of decorations (herbs and branches that look like wings, round fruits placed in the mouth or in the eyes, etc.). Such dressing is a symbol of pantagruelic grandeur, and perhaps, in its own way, a religious metaphor of the sacrifice to the god/s. How can people, in such a case, omit to consider the connection with the living, sentient being that once that grotesque display was? Actually, in this case the connection is not omitted (it could not be clearer, in fact) but still a metalevel is introduced—and that is the metalevel of anthropocracy at its boldest: instead of not wanting to know what meat was before becoming meat, now people will put a special stress on this, reminding themselves of their dominance over other animals and their God-given power over nature.

10.4 Conclusions

The essay was written with an intention to contribute to the debate veganism-carnism via the following claims:

(1) The legitimization of the word "carnism" as the real terminological and conceptual counterpart of the word "veganism", due to the complete circle of cultural and ideological implications that the former (in Melanie Joy's formulation) offers. In particular, I suggest to renounce the terms "carnivore" and "omnivore" (both inaccurate scientifically, with the latter more appropriately replaced by "polyfagous" or "eurifagous").

(2) The argument that carnism, like any other form of anthropization, is filtered via three similar-but-compatible mental attitudes: anthropocentrism, speciesism and anthropocracy. In this essay I have also attempted a definition and a distinction of the three terms, arguably using the concept of "anthropocracy" for the first time.

(3) An understanding of anthrozoological relations (marked by any of the three above-mentioned filters) as intergroup biases, in the sense meant by social psychologists like Allport or Tajfel.

(4) A rejection the expression "cognitive dissociation", as applied to the apparent contradiction between showing sympathy/compassion for one species and contempt/indifference for another. In this essay I have argued that one should rather refer to this attitude as different "metalevels of representation".

(5) This concept, among other things, allows to focus on the communication strategies operated by carnism to convey or disguise (depending on the case) the kind of information that veganism is ethically pointing the finger at: the promotion of meat as product, necessity, display of luxury, nutritionally-valuable aliment, and so forth. This brought to the identification of four communicative spaces: Hidden/disguised, Ambiguous/antithetic, Clear/informative, Exposed/Ostentated.

References

Allport, Gordon W. 1954. *The nature of prejudice*. Reading, MA: Addison Wesley.

Bartolommei, Sergio. 1995. *Etica e natura*. Roma-Bari: Laterza.

Berkmanienė, Aušra, and Dario Martinelli. 2018 forthcoming. The politics and demographics of veganism. *International Journal of Semiotics of Law*.

Celentano, Marco, and Dario Martinelli. 2018 forthcoming. Ethology of the freed animal: Concept, methods, projects. *Behavioural Sciences*.

Freston, Kathy. 2011. *Veganist*. New York: Weinstein Publishing.

Iacobbo, Karen, and Michael Iacobbo. 2006. *Vegetarians and vegans in America today*. Santa Barbara: Greenwood Publishing Group.

Joy, Melanie. 2001. From carnivore to carnist: Liberating the language of meat. *Satya* 8 (2): 26–27.

Joy, Melanie. 2010. *Why we love dogs, eat pigs, and wear cows: An introduction to carnism*. San Francisco: Conari Press.

Martinelli, Dario. 2010. *A critical companion to Zoosemiotics: People, paths, ideas*. Berlin-New York: Springer.

Martinelli, Dario. 2016. *Arts and humanities in progress: A manifesto of numanities*. Berlin-New York: Springer.

Maurer, Donna. 2002. *Vegetarianism: Movement or moment?*. Philadelphia: Temple University Press.

Singer, Peter. 1975. *Animal liberation: A new ethics for our treatment of animals*. New York: HarperCollins.

Smith, Andrew F. 2013. *Eating history*. New York: Columbia University Press.

Tajfel, Henri. 1981. *Human groups and social categories*. Cambridge: Cambridge University Press.

Chapter 11
Translation of Culture-Specific Items in Menus

Saule Petronienė, Jolita Murauskaitė and Jolita Horbačauskienė

Abstract Translation is a difficult phenomenon because it is concerned not only with a pair of languages but also with the cultural aspects that languages are related to. One of these aspects consists of culture-specific items (CSIs)—material or spiritual concepts that are present in one culture and are absent in another. In the 21st century, because of globalization, people encounter CSIs on a daily basis. Therefore, it is very important to understand the concept of CSIs and the impact they have, along with the misunderstandings and problems they may cause. Thus, the object of this paper is CSIs and their translation in Lithuanian menus. The aim of the paper is to analyze translation of CSIs in the menus of Lithuanian restaurants. The analysis of CSIs is performed according to the translation strategies introduced by Davies (Trans 9(1):65–100, 2003). The results of the analysis demonstrate that strategies which tend to keep the content of the original CSI or neutralize/explain cultural references are often used to translate Lithuanian CSIs, whereas translation strategies that preserve the original form of CSIs or tend to replace one cultural reference with another are more often used to translate foreign CSIs. Since CSIs may not have equivalents in other languages, cases when few strategies are employed to translate the same CSI appeared as well as some mistakes in translation.

Keywords Culture-specific item (CSI) · Translation · Menus
Translation of CSIs · Translation of menus

S. Petronienė (✉) · J. Murauskaitė · J. Horbačauskienė
Kaunas University of Technology, Kaunas, Lithuania
e-mail: saule.petroniene@ktu.lt

J. Murauskaitė
e-mail: jolita.murauskaite@gmail.com

J. Horbačauskienė
e-mail: jolita.horbacauskiene@ktu.lt

A. Olteanu et al. (eds.), *Meanings & Co.*, Numanities - Arts and Humanities in Progress 6, https://doi.org/10.1007/978-3-319-91986-7_11

11.1 Introduction

Nowadays translation is not considered as simple transference of words from one language to another any more since context matters as much as the text itself, meaning that the culture-related aspects cannot be excluded. One of the cultural peculiarities that translators deal with is the culture-specific items (CSIs)—material or spiritual concepts, which occur in one nation, country or region and do not exist in another, due to the different development of culture and language. Their translation imposes a great many problems for translators (Aixela 1996; Newmark 1988), such as a lack of equivalents (Nord 1997; Nababan 2008; Petrulionė 2012), lexical plus connotative meanings (which may vary depending on the context) (Bassnett 2002; Balčiūnienė 2005), loss or addition of information (Nababan 2008), and different grammar rules (e.g. rules for capitalization). Besides, translators need to pay additional attention to the source language (SL) and target language (TL), in addition to both cultures and their backgrounds (Maksvytytė 2012), while considering the authors and readers, along with their expectations when translating CSIs.

In the 21st century, translators are not the only ones who deal with CSIs, mainly due to advancing globalization, increased access to communication means and easier traveling options. Therefore, cultures are in contact with each other more often, and people encounter CSIs on a daily basis. Hence it is very important to understand the concept of CSIs and the impact they have, and the misunderstandings and problems they may cause. Thus, the object of this paper is CSIs and their translation in cuisine menus. The study aims at exploring translation of CSIs in cuisine menus: presenting the translation problems and defining translation strategies for CSI, discussing results of the analyzed data as well as comparing the translation strategies used for Lithuanian and foreign cuisine CSIs translation.

11.2 Theoretical Background

Culture plays a major role in what concerns the formation of our thinking, behaving, and, altogether, of our perception of reality. Every culture contains concepts for describing their traditions, habits, social status, measurement units, etc. In every culture these concepts are perceived differently and when two or more cultures come into contact, all the differences suddenly have a huge importance and consequence. This is when the phenomenon of culture-specific items appears.

11.2.1 The Concept of CSI

Scholarship has not come to an agreement in what regards a term to name that which is known to one culture and foreign to another, whether referring to a word, an object, or a concept. Such entities are termed "culture specific, culture bound references/elements/terms/items/expressions, realia, allusions, [...] or cultural references" (Ranzato 2013, p. 47). The term "culture-specific item" is used in this paper. They are perceived as the objects that appear in one culture and are foreign or understood differently in another. Taking into account CSI authenticity and the difficulties while translating them, Aixela proposes that CSIs are "those textually actualized items whose function and connotations in a source text (ST) involve a translation problem in their transference to a target text (TT)" (Aixela 1996, p. 58). However, this definition is a little too narrow because as Braçaj (2015) points out, "cultural elements do not involve just the items" (p. 476). According to Kvašytė (2010), CSIs may also be spiritual cultural elements. In addition, CSIs may be both verbal and non-verbal signs (Ramière 2006) and are words that define cultural concepts, items and expressions common for one culture, whilst conveying national, local and historical background (Balčiūnienė 2005).

For achieving a more insightful understanding of CSIs, a classification is needed. Many attempts have been made to categorise CSIs, however, the classification of them varies from one author to another. These classifications have been created according to different criteria: either the semantic meaning of the CSIs or the origin from which culture the CSIs derive. The first type of classification is proposed by Newmark (1988) who divides CSIs according to their meaning, e.g. ecology; material culture; social culture; organizations, customs, activities, procedures, concepts; gestures and habits (Newmark 1988). Food is placed into the category of material culture by Newmark.

Almost exactly the same classification of CSIs is proposed by Balčiūnienė and Mikutytė; they share many similar categories (geographic and ethnographic terms, folklore and mythological terms, household terms and public historical terms). However, they place food into different categories, i.e. Balčiūnienė places food under the heading household terms, while Mikutytė puts the category of food under the heading ethnographic terms, even though they both share the category of ethnographic terms. This division again proves that there is a lack of consistency in translation-related terms, in addition to being vaguely defined.

A very different classification of CSIs is proposed by Ranzato (2013): realistic references (source culture references, intercultural references, third culture references, target culture references) and intertextual references (overt intertextual allusions, covert intertextual allusions, intertextual macroallusions). First of all, Ranzato categorizes CSIs from the point of view of the TL audience and its relationship to the SL text, and according to the origin from which culture CSIs derive, not semantic meaning. Realistic references are non-fictional persons, objects and events known to a particular audience; intertextual references are allusions to the text which create the bond between the texts (Ranzato 2013). According to Ranzato, CSIs of food belong to the category of realistic references.

11.2.2 Problems of CSIs Translation

Newmark (1988) divides difficulties of translation of words into two problems and one of them covers CSIs: "you find them hard to translate" (p. 33), meaning CSIs impose many problems translators have to deal with.

First of all, cultures and languages develop differently, thus, there are no two languages that would be similar enough to represent the same reality (Bassnett 2002). The impossibility of precise equivalence stems from semantic, socio-cultural and grammatical differences between the source and the target language (Nababan 2008). In the endeavour of translating CSIs, it is important to acknowledge that perfect equivalence is not a universal virtue in translation in general. While there is scholarship confronting the equivalence of meaning as the core issue to translating words with cultural implications (Petrulione 2012), there is a stronger case advocating that semantic equivalence does not underpin cultural translation (Nord 1997). Nord explains this suggestion within the example of a Brazilian school report used to apply to a German higher education institute, which will not become a German "Abiturzeugnis" by being translated into the German language (p. 141), it will remain a Brazilian School Report. Such a kind of translation is not equivalent to the original and there is no need to insist that there be one. Therefore, a question that needs to be addressed is whether equivalence in the translation of CSIs needs to be sought after at all.

Furthermore, CSIs not only carry linguistic information but also "the whole set of extra-linguistic criteria" (Bassnett 2002, p. 22) such as cultural, historical and ethnographic background. Therefore, translators need linguistic competence as well as awareness in that specific culture or knowledge about its CSIs in order to create translations (Maksvytytė 2012). Consequently, omission and addition of information cannot be avoided (Nababan 2008). Therefore, the problem which arises consists in considering what information should be kept and what should not. What can be considered as unnecessary can be omitted. It is important to note, that the main goal of a translator is to make the text coherent and understandable; either the author is brought to the reader or the reader to the author, but this is secondary as long as the main ideas are successfully conveyed.

Moreover, CSIs may have different meanings in various fields, e.g. historical, geographical and political (Balčiūnienė 2005). TL readers may not be able to understand CSIs in every/different context (Pažūsis 2014, p. 244), thus, translators cannot limit translation only to the frames of written text, for example; discourse also matters. Furthermore, CSIs may cause grammatical problems, e.g. whether to capitalize each word of a CSI as it is written in the original text or not, considering that the rules of the TL may dictate the spelling in lowercase letters.

Finally, the last issue translators face is accountability. What is the role of author, client and reader? Does the pair of languages and gap between the cultures matter? Additionally, socio-cultural factors have to be taken into account, but the question remains of how much they should influence/are influencing the process of translation.

11.2.3 Strategies for CSIs Translation

Because CSIs are complicated phenomena, many translation strategies were introduced solely for their translation. The majority of these classifications follow a "common general progression from the most exoticizing to the most domesticating" (Ramière 2006, p. 156). Foreignization (exoticizing) is the strategy tending to retain information from the ST. Nord (2006) states, it is a "strategy of keeping close to the source text" (p. 56). By employing this strategy, the cultural and linguistic differences of two languages are being saved. Domestication, on the other hand, is the strategy of making the ST as familiar to TL readers as possible; by adapting "target text to target-culture conditions in order to make the source-text author's intention or purpose work for the target audience" (Nord 2006, p. 56). The purpose and rationale of this strategy is to make the text fluent and smooth as if it is not a translation. Such a progression from the most to the least amount of foreignizing is very common and many scholars have discussed and classified strategies according to it. Thus, the names under which foreignization and domestication are widely known from one author to another (Nord 2006). Foreignization is named formal equivalence (Nida 1964), substitution (Aixela 1996) and retrospective/source-oriented (Nord 2006); domestication is also introduced as dynamic equivalence (Nida 1964), conservation (Aixela 1996) and prospective/target-oriented (Nord 2006).

The most foreignizing strategies are preservation of form or content. Preservation of form means that the original form of the CSI is preserved in the TT and is employed either when the CSI has no close equivalent in the TL (Davies 2003) or translators tend to retain as much as possible of the original reference (Aixela 1996), or the CSI is thought to be totally clear and easy to understand for the TL readers (Balčiūnienė 2005). Preservation of content means that the CSI is translated by copying its semantic or derivational motivation with the means of the TL. This strategy is used to maintain compactness and semantic clarity between SL and TL CSIs (Pažūsis 2014); when the word contains an easily recognizable descriptive element (Davies 2003); and "to preserve any wordplay" (Ranzato 2013, p. 105), because every word of the CSI receives literal translation.

Another translation strategy consists of the addition of words (Davies 2003) when the original CSI is maintained and further explanation is provided. According to Pedersen (2005), "by using this strategy, the translator intervenes to give guidance to the TC audience" (p. 5), meaning additional information is added to clarify the message of the CSI. The advantage of this strategy is that in most cases there is no need to add any further meta-comments, and the translation reflects all the aspects of the SL CSI. However, the provided additional information may become too detailed and complicated (Pažūsis 2014). Thus, it is important for translators to find a balance. This strategy can be employed in two different ways: information is added inside the text (addition inside the text) or outside the text in the form of footnotes, glossaries (addition outside the text).

Furthermore, halfway between the most foreignizing and the most naturalizing translation strategies there is globalization (Davies 2003). As Paluszkiewicz-Misiaczek (2005) explains, globalization consists in replacing a specific word/concept with a more commonly known, general one. By employing this strategy, the foreign content and semantic shades of the original CSI are neutralized (Pažūsis 2014) and the text becomes accessible to audiences from a wider range of cultures (Davies 2003).

Contrary to globalization, the localization strategy aims to find another cultural reference that would be familiar to the TL reader. According to Pedersen (2005), this strategy involves deleting the original CSI and replacing it with another one that is expected to be known to the TL readers. Localization is used when the information carried by the original CSI is not particularly important, because the latter is neutralized as well as assimilated with the meaning of the TL CSI. What matters in this case is to induce the same feelings, emotions or associations as the original CSI was supposed to do (Pažūsis 2014). This translation strategy is viewed from two different angles. Firstly, it refers to localization as rendering a more general concept into a more specific one (Molina and Albir 2002; Paluszkiewicz-Misiaczek 2005; Ranzato 2013); while the second opinion interprets localization as the SL CSI substitution with the local CSI with the same meaning (Aixela 1996; Molina and Albir 2002; Davies 2003; Balčiūnienė 2005; Pedersen 2005; Leonavičienė 2011; Schmidt 2013).

One of the most domesticating strategies is transformation (Davies 2003). By employing this strategy, the CSI meaning is modified—the SL CSI is replaced with a cultural reference of the TL that the readers "can identify and which is easy to understand, familiar and appealing to them" (Paluszkiewicz-Misiaczek 2005, p. 244). Even though the content of the original CSI is distorted, it is likely that the new one will have the same impact on the TL readers (ibid). Localization and transformation can be confused, as they both involve changing the original CSI with another one. The difference is that the CSI translated with the strategy of transformation may have a different meaning from the original CSI but the same effect, while localization has both the same meaning and effect.

Omission (Davies 2003; Paluszkiewicz-Misiaczek 2005; Pedersen 2005; Leonavičienė 2011) is considered to be the most drastic of all strategies for CSIs translation. Omission is a translation strategy by which the original CSI which is "present in the ST, is not transferred to the TT and disappears" (Ranzato 2013, p. 111). This strategy is used when the CSI is extremely difficult to translate (Paluszkiewicz-Misiaczek 2005), when the CSI is considered to be not relevant nor important (Balčiūnienė 2005), or to avoid redundancy (Ranzato 2013).

The last translation strategy to be discussed is creation (Davies 2003). This strategy is used when the CSI is created in the TL text and is not present in the original text. This strategy is employed when the CSI is considered to be interesting for the TL readers (Aixela 1996) or the text would sound rather natural and coherent if the CSI is introduced. Creation is "subjected and dictated by the personal taste of the translator" (Ranzato 2013, p. 113), because it is up to the translator to choose whether to introduce the new CSI or not. According to Aixela (1996), creation is very seldom used.

Summing up, there are many translation strategies introduced for CSIs translation. The number of strategies varies from one author to another, as well as the names and definitions of them.

11.3 Methodology

Nowadays, CSIs are more common than they have ever been. People struggle every day with cultural references and the misunderstandings caused by them. The translation of CSIs has been researched and discussed by a number of scholars (Aixela 1996; Nord 1997; Bassnett 2002; Davies 2003; Mikutytė 2005; Pedersen 2005; Kvašytė 2010; Leonavičienė 2011; Maksvytytė 2012; Braçaj 2015). Since it is very important to have a clear idea of what CSIs are, what impact they have on daily life and which translation strategies may be used in one or another case. However, to the best of our knowledge, no research on the translation of CSIs in menus has been conducted.

The definition of CSIs used in this paper is introduced by Aixela (1996), who claims that cultural-specific items refer to words, phrases and idiomatic expressions that come from one culture and have no equivalents in another; they may be abstract or concrete, material or spiritual, from religious belief to the type of food; thus, CSIs cause translation problems because of their functions and meanings.

The analysis of CSIs has been done according to the translation strategies introduced by the scholar Davies (2003): preservation, addition, globalization, localization, transformation, omission and creation. Descriptive content analysis and statistical methods have been used. Examples of CSIs and their translation to English have been selected from 15 menus of Lithuanian restaurants.

CSIs of Lithuanian and foreign cuisine have been extracted in order to analyze their translation and to compare the translation strategies that have been used. In total, 536 examples have been found: 184 examples of Lithuanian cuisine CSIs and 352 examples of foreign cuisine CSIs.

11.4 Analysis of CSIs Translation in Lithuanian Menus

Nowadays, CSIs are widely spread and commonly used in many different areas. First of all, translators need to have knowledge of both source and target cultures and languages, because the differences between them are the reason why CSIs appear in the first place. Translators have to decide whether to foreignise or domesticate the text and choose the strategy (or a combination of them) for CSIs translation. The focus of the present research is the choice of translation strategies that have been used to translate both Lithuanian and foreign cuisine CSIs.

The most foreignizing translation strategy is preservation, when either the form or the content of the original CSI is retained in TL text. It is divided into two subcategories: preservation of form and preservation of content. The first subcategory is transference of the CSI to the TL text in its original form. It is used when the CSI does not have a close equivalent in the TL or the translator tends to keep the foreign sounding. This strategy is much more often used to translate foreign CSIs, because they are supposed to be better known, their forms are considered to be recognizable and, thus, are retained in the target texts. For example, "*Canelone*"–*Canelone; karpačas—carpaccio*. The only difference stands in whether the CSI is orthographically adapted or not.

Lithuanian CSIs, on the other hand, are not very well known for the target audience (tourists) nor are they widely spread, thus, preservation of content is used instead. For example, *bulvių blynai—potato pancakes, raugintų kopūstų sriuba—sauerkraut soup*. This tendency to retain either the form or content may be explained by popularity and the fact of how well the CSIs are known to readers, which depends on the dominance of SL culture, i.e. the more dominant the culture, the more widely its cultural references are spread, and the more options to keep the original forms exist. The second reason to choose either preservation of form or content is the gap between cultures—the size of the gap is directly related to how widely the CSIs are going to be spread in one culture or another.

For the same reasons, the strategies that tend to explain/neutralise cultural references, addition and globalization, are more often used to translate Lithuanian CSIs. Addition preserves the original CSI, i.e. it places emphasis on the SL cultural reference and gives additional information for the reader in order to understand the meaning better. Consider the examples: *Vėdarai*—"*Vėdarai*" (Potato sausages); Skilandis—*Lithuanian sausage* "*Skilandis*". Addition may be used inside or outside the text. No examples of the latter have been found. Addition inside the text means the explanation is written in line with the text. In some cases, it is placed in brackets to focus more on the CSI. In general, additional information explains either the method of preparation, names the products of which the dishes are made of or defines what the CSI is itself.

Globalization tends to replace the original word/concept with a more general and neutral one, consider: *Žemaitiška grybienė—Samagitian mushroom soup; Silkė „patale"—Herring with vegetables*. The first example shows how the particular name of the Lithuanian soup *grybienė* is generalized into *mushroom soup*. For SL readers it is clear what the dish is made of (*grybai—mushrooms*) and that *grybienė* is a soup, thus, the extra word *sriuba* (*soup*) is not necessary. It would make the text overfilled with additional words, sounding unnatural or even illogical. In the second example, the word "*patale*" is generalized into the products of which the dish is made of. For SL readers it is obvious that *Silkė „patale*" is a herring covered with various types of vegetables while for the TL audience it would not make any sense. Thus, the translator chooses rather to explain/generalise what the original CSI means than to keep the form or the content.

Localization, which tends to replace one cultural reference with another, is more often used to translate foreign CSIs because it is easier to find their cultural equivalents. Localization may be employed in two ways. Firstly, it is when a general word is replaced with a more specific concept, e.g. *Austriškas pyragas—Strudel*. *Austriškas pyragas* means 'Austrian cake' which is a very general concept; any pie that is made in Austria could be called this. However, translation is *Strudel*, which means a specific kind of pie having its own preparation way and ingredients. The second type of localization is changing one CSI with another, e.g. *Lietinis blynas—Crepe*. *Lietinis blynas* is known to the SL audience as a thin pancake, made of flour, cooked in a frying pan and served with either sweet or savory filling. The translator introduces the foreign CSI *crepe*, which has exactly the same meaning. Using localization ensures that the meaning is absolutely transferred. The text (CSI) sounds natural because TA is familiar with the introduced CSIs.

Other strategies (transformation, omission and creation) are seldom used to translate both Lithuanian and foreign CSIs, because translators tend to avoid altering or deleting the original message of the CSIs or introducing a new cultural reference in the menus. Even though the translation of menus seems relatively loose and without strict frames, translators restrain themselves from drastic changes in order to avoid misunderstandings and misinterpretations.

Sometimes translators employ a combination of a few different strategies to translate the same CSI, since there is no set of rules of saying which strategy should be employed in which case. Consequently, the same CSI can be translated by using different strategies. Moreover, because of the lack of consistency in defining translation strategies and how they should be employed, such cases occur when the same CSI is translated differently, even though the strategy that is used to translate it is the same.

In order to make the translation as good and naturally sounding as possible for the TL readers, translators sometimes employ a few different strategies when dealing with translation of CSIs. For example, Lithuanian CSI *Didžkukuliai—Potato dumplings* is translated using *localization* and *addition*. First, the original CSI *didžkukuliai* is localized into *dumplings* (both dishes can be made of potatoes and include a filling of meat, curd or vegetables), i.e. one cultural reference is replaced with another that is easier to understand for the TL readers. To provide even more clarity, the translator uses *addition*. Dumplings can be based on potatoes, flour or even bread, while *cepelinai* is only made from potatoes. To avoid misunderstanding, the translator adds an extra explanation, which says what the dish is made of (*potatoes*). Foreign CSI *Baklažanų užkepėlė "Parmigiana"*—*Parmigiana* is translated using *omission* and *preservation of form*. First of all, the original form is preserved (*parmigiana*) and then omission applied by expecting that the TA is familiar with the CSI and does not need any other information except the title of the dish.

Another interesting tendency that has been noticed is the same CSI translated with the same strategy can result in various expressions. The Lithuanian dish *Balandėliai* is translated in three different ways: *Minced meat stuffed and stewed cabbage leaves, Stuffed cabbage rolls, Meat-stuffed cabbage rolls*. All the examples are translated by using addition inside the text without the original CSI. Even though the strategy is the same, the translation is different—as different additional information is provided. Similar cases occurred with foreign CSI translation. *Rikotos sūris* has been translated as *ricotta cheese* and *cottage cheese*. In the first example, the original form is preserved whereas in the second example the same CSI is globalized into cottage cheese. Globalization is therefore used to make the translated text sound more neutral. The original CSI may not be that widely known, while the general name of the food is understood everywhere by everyone. Thus, in the second case the translator decides not to risk and globalizes *ricotta* into *cottage cheese*. All of the strategies are employed in the translation of a menu. However, different trends on which strategies or combinations of them are used to translate Lithuanian and foreign CSIs are noticed.

Increasing multicultural environment inevitably enhances the translation of CSIs, thus, this particular area of translation is becoming more significant. Therefore, the results of the research could be helpful for the translation of CSIs; also, they may be considered by other researchers as guidelines for further investigation.

11.5 Conclusions

The results of the analysis demonstrate that the most common strategy for the translation of Lithuanian CSIs is preservation of content, while preservation of form is used for the translation of the majority of foreign CSIs. The latter distribution of the translation strategies suggests that Lithuanian CSIs are not very well known for the target audience (tourists) nor are they widely spread and, thus, the content instead of the form is preserved in the translated texts. On the other hand, foreign CSIs are supposed to be better known, their forms are considered to be recognizable and, thus, are retained in the target texts. For the same reasons, the strategies that tend to neutralize/explain cultural references (globalization, addition) are more often used to translate Lithuanian CSIs, while localization that tends to replace one cultural reference with another is more often used to translate foreign CSIs, mainly because it is easier to find their cultural equivalents. Transformation, omission and creation are seldom used to translate both Lithuanian and foreign CSIs, because translators tend to avoid altering or deleting the original message of the CSIs or introducing a new cultural reference in the menus.

In translating CSIs, translation strategies are specifically employed, as the analysed data suggests. This use of strategies is due to the CSIs not having equivalents in languages other than the one to which they belong. Consequently, a combination of strategies to translate one CSI has been found in the translation of menus. Taking into account that there is no set of rules for which strategy has to be used and when they should be used, cases occurred when the same CSI has been translated by using different strategies in translation of both Lithuanian and foreign cuisine CSIs.

Also, it has to be noted that the researched material is highly specialized because of the nature of the texts, which influences the use of translation strategies. The translation of CSIs is likely to be influenced by other contexts and material, and to another extent and/or in another manner. Thus, due to the specificity of the researched material and the fact that no research on translation of CSIs in menus has been found, the analysis (and comparison) of translation of CSIs in other texts and contexts was not purposeful.

References

Aixela, J. F. 1996. *Culture-specific items in translation.* Translation, Power, Subversion. Clevedon: Multilingual Matters.

Balčiūnienė, I. 2005. *Apie realijų vertimą. Lietuvos literatūros vertėjų sąjunga.* Retrieved from http://www.llvs.lt/?recensions=34. Accessed 01 Mar 2016.

Bassnett, S. 2002. *Translation studies*, 3rd ed. New York, NY: Routledge.

Braçaj, M. 2015. Procedures of translating culture-specific concepts. *Mediterranean Journal of Social Sciences* 6 (1): 476–480.

Davies, E.E. 2003. A goblin or a dirty nose? the treatment of culture-specific references in translations of harry potter book. *The Translator* 9 (1): 65–100.

Kvašytė, R. 2010. Pasvarstymai apie beekvivalentę leksiką, arba latvių Joninės lietuviškai. *Gimtojia kalba* 10: 11–19.

Leonavičienė, A. 2011. Lietuvių kultūrinių teksto reikšmių interpretacija ir vertimas. *Studies about languages* 19: 39–44.

Maksvytytė, J. 2012. Sąvokų realija ir realijos pavadinimas apibrėžimo problema. *Studies about Languages* 21: 50–56.

Mikutytė, J. 2005. Realijų rūšys ir vertimo būdai. *Lietuvos literatūros vertėjų sąjunga.* Retrieved from http://www.llvs.lt/?recensions=29. Accessed 22 Feb 2016.

Molina, L., and A.H. Albir. 2002. Translation techniques revisited: a dynamic and functionalist approach. *Translators' Journal* 47 (4): 498–512.

Nababan, M. R. 2008. *Equivalence in translation: Some problem-solving strategies* (Doctoral dissertation). Retrieved from Translation Theory http://www.proz.com/translation-articles/articles/2071/1/EQUIVALENCE-IN-TRANSLATION%3A–SOME-PROBLEM-SOLVING-STRATEGIES. Accessed 15 Mar 2016.

Newmark, P. 1988. *A textbook of translation.* Prentice Hall International (UK) Ltd: New York & London.

Nida, E.A. 1964. *Toward a science of translating. With special reference to principles and procedures involved in bible translating.* Leiden: Brill.

Nord, C. 1997. *Translating as a purposeful activity: Functionalist approaches explained (translation theories explained)*. New York, NY: Routledge.

Nord, C. 2006. Translating for communicative purposes across culture boundaries. *Journal of Translation Studies* 9 (1): 43–60.

Paluszkiewicz-Misiaczek, M. 2005. Strategies and Methods in Dealing with Culture Specific Expressions on the Basis of Polish-English Translations of Certain Administrative and Institutional Terms. *Thepes* 3: 243–248. Retrieved from http://www.phil.muni.cz/plonedata/.

http://www.wkaa/Offprints%20THEPES%203/TPES%203%20(243-248)%20Paluszkiewicz.pdf. Accessed 15 Mar 2016.

Pažūsis, L. 2014. *Kalba ir vertimas*. Vilnius: Vilniaus universitetas.

Pedersen, J. 2005. How is Culture Rendered in Subtitles? *Challenges of Multidimensional Translation: Conference preceding*. Retrieved from http://www.euroconferences.info/proceedings/2005_Proceedings/2005_Pedersen_Jan.pdf. Accessed 15 Mar 2016.

Petrulionė, L. 2012. Translation of culture-specific items from english into lithuanian: The case of Joanne Harris's novels. *Studies about Languages* 21: 43–49.

Ramière, N. 2006. Reaching a foreign audience: Cultural transfers in audiovisual translation. *The Journal of Specialised Translation* 6: 152–166.

Ranzato, I. 2013. *The Translation of Cultural References in the Dubbing of Television Series* (Unpublished doctoral dissertation). Translation Studies Unit, Imperial College London, London.

Schmidt, G. 2013. Foreignization and domestication in the Croatian translations of Oscar Wilde's "The Picture of Dorian Gray". *Jezikoslovlje* 14 (2–3): 537–548.

Chapter 12
Semiotic Practices in TV Debates

Nicolae-Sorin Drăgan

Abstract The purpose of this paper is to propose a methodological framework to analyze the dynamics of the positioning acts of political actors in dialogic forms of interaction, such as final TV debates for presidential elections, from the perspective of positioning theory (Davies and Harré 1990; Harré and Gillet 1994; Harré and van Langenhove 1998; Moghaddam et al. 2008; Moghaddam and Harré 2010; Harré and Moghaddam 2016). Positioning theory activates a tridimensional semiotic model, which includes *positioning acts*—semiotic configurations with a certain stability in the communicational practices of the community—*storylines* and the implicit *illocutionary force* of any discursive exchange. This paper uses an interdisciplinary approach—analysing the positioning complementarily with the Eudico Linguistic Annotator (henceforth ELAN) framework for multimodality research—to capture on the one hand the dynamics of the discursive exchange in the episode we considered, and on the other hand to disambiguate the positioning (and the interpretation thereof) of the political actors involved in the debate.

12.1 Introduction

In this paper we analyze the way in which social actors are placed and act discursively (strategic positioning) in one of the most important forms of political communication, namely final debates for presidential elections.

During the labour of sign production (communication semiotics) (Eco 1976), or of the significant practice understood as the "meaning production process" (Kristeva 1980: 251), any social actor "builds/takes a position (identity) in relation to their interlocutors" (Beciu 2009: 35). For example, at the last presidential debate in Romania, broadcasted by B1TV on the 12th of November 2014, challenger Klaus Iohannis, leader of the Christian Liberal Alliance (CLA), said:

N.-S. Drăgan (✉)
National University of Political Studies and Public Administration (NUPSPA), Bucharest, Romania
e-mail: sorin.dragan@comunicare.ro

A. Olteanu et al. (eds.), *Meanings & Co.*, Numanities - Arts and Humanities in Progress 6, https://doi.org/10.1007/978-3-319-91986-7_12

> [...] given my training as a teacher, I think that it is a problem when a child plagiarizes. When somebody who writes his doctoral dissertation cheats, it is a proportionally greater issue and it is called plagiarism. Anywhere in the world, anyone among the politicians who is caught plagiarizing is forced by the political class to quit politics. In our country, one of these plagiarists wishes to become Romania's president. (O'Clock Nov. 13, 2014)

Through this statement, Klaus Iohannis—currently president of Romania—has not just made an offensive statement (attack) against the incumbent Prime Minister Victor Ponta, leader of the Social Democratic Party (SDP). Iohannis performed an *act of positioning*. Practically, he indexed a certain type of social relationship (teacher/student). More recently, during the final TV debate on May 3, 2017, French presidential candidate Marine Le Pen did the same thing when she said: "Mr. Macron... I see you're trying to play the teacher and pupil with me" (BBC May 4, 2017). In the same TV debate candidate Emmanuel Macron—who finally won the French presidential election—portrayed his opponent in the presidential race as "The grand priestess of fear".

The replies of political actors are not mere verbal statements, but meaningful connections that are not visible at the apparent surface of the text. In any form of discursive interaction "words, expressions, propositions, etc., change their meaning according to the *positions* held by those who use them" (Pêcheux 1982: 111; cited in Rovenţa-Frumuşani 2012: 149). We shall see that such semiotic practices—as the positioning acts in this article are understood—activate certain "power" relations that are pre-established in the social memory. The positioning acts "necessarily involve relations of power" (Harré and Gillett 1994: 123). Such relations "set the social identity, roles and style of communication of the protagonists" (Rovența-Frumușani 2012: 22). In this way, some representations of situations and social relations are updated.

As illustrated in previous examples, a discourse signals a position to-wards the content of communication (what it enunciates), as well as a (relative) position towards the participants to the discursive act. Therefore, in communication acts, social actors communicate not just content, but also the way they relate to that content (positioning). Thus, positioning becomes individual practices in discursive behaviour, which can configure a certain communication style. In the perspective that discourses form a "semiotic web" (Rovența-Frumușani 1999: 32) the practice of positioning can be regarded as an effect of intertextuality of discourse. This idea is further developed in the third section of this article.

In the next section, I briefly describe the televised debate as electoral practice and the essential experience of electoral communication. In the third section I present the positioning theory and discuss positioning acts as semiotic practices in TV debates. In the fourth section I propose an analytical instrument built on an interdisciplinary approach—analysing the positioning complementarily with a professional framework for multimodality research, ELAN—to analyze the dynamics of the positioning acts of political actors in TV debates. In the last part of the paper (Sect. 12.5) I conclude.

12.2 TV Debate, Between Electoral Practice and the Experience of Dialogue

Today, public space has turned into a place of "narcissistic exhibitionism" (Bourdieu 1998: 14) for political actors. According to Pleşu (2010), the place of authentic dialogue was taken by the "mystique of dialogue".[1] In this context, it is necessary to understand the dialogue as opening towards the other (Pleşu 2010). The appropriate solution would be understanding communication in Constantin Noica's terms, as "Eucharist, because it means communion, it means a participation of each of us to the thoughts and to what the Other has to say", as Andrei Pleşu explained in a 2015 televised interview for the Romanian public TV channel. A consistent debate needs a authentic, genuine dialogue. Political actors cannot waste the experience of dialogue in TV debates when they want to make sense of the relationship with others.

The televised debates is considered "superior to other communication forms" (Pfau 2002: 251) and can be regarded as a special form of communication that reveals, on the one hand, the relationship of social actors with the language, and on the other hand, it expresses their capacity of understanding otherness.

In this political and media context we are witnessing a shift in the shapes of political discourse towards a show-type politics or media events (McKinney and Carlin 2004: 214; Beciu 2009: 116). In such a paradigm of current political communication, *the act of observation* tends to replace the *act of participation* (Erickson 1998: 145). This is one of the consequences of political discourse turned into a show.[2]

According to Lempert and Silverstein (2012), in the televised debate, messages are intentionally made theatrical, in the sense of constructing a certain moral profile of the candidate. Thus, the political message becomes a *personal brand*, giving political actors a kind of "biographical aura" (Lempert and Silverstein 2012: 100). The construction of the *political character* involves a huge semiotic work. The media exploits the theatricality and show appearance of the televised debate, giving

---

[1]It is politically correct to say: "Everything is solved by dialogue"; "Dialogue is holy. Any problem, any crisis can be solved by talking. Dialogue"; "Everything is dialogue, tolerance, opening, dialogue etc" This mythology of dialogue, which has its good reasons, should not be idolised, though. Dialogue is not possible in whatever conditions, and we, in Romania, today, should know it too well. Dialogue needs a certain type of situation to be born lawfully and to be carried out coherently. One cannot have a dialogue with a wall; one cannot have a dialogue with something closed. Dialogue means, on the contrary, availability, opening. Closing is external to the idea of dialogue.

[2]In such a paradigm the media rather values the image of political actors to the detriment of the content of messages (Postman 1985; Hellweg et al. 1992: 79). Erickson (2000) calls the media propensity to image to the detriment of the content of messages, the "*visual turn*" in presidential rhetoric. Eco (1976) sees in this reality a border between *neo-television* preoccupied with *personalization* of social actors, public image and relations to public, and *paleo-television*, focused on the content of messages, the audience being positioned in a generic reference (Beciu 2009: 109), without interactional valences.

the illusion of living a *Pirandelian-type* experience. Viewers try to figure out the subtle distinction between actors and characters, between features and masks. Who is the proper performer of the president part? They may even get to decide in favour of one of the actors, in other words, they will have a say in the way the play is constructed. From here stems the illusion of a *theater within a theater*. Viewers are caught in the subtle difference between appearance and reality.

Furthermore, McKinney and Lamorueux (1999) discuss the narrative structure of debates, which are built like a "media drama in three acts" (as cited in McKinney and Carlin 2004: 214–215). The first act starts with the *debate about debate* scene. In Romania, for instance, starting with the presidential election campaign in 2004, the topic of election debates in media has arguably become "an election theme and a criterion for assessing the actions of candidates during their campaigns" (Beciu 2015: 254). The first act of the drama is where the stage is set. One can notice in the public space that people are talking about the uncertainty of the event, about the manner in which the debate will be broadcast, about the structure of the debate etc. (McKinney and Carlin 2004: 214). The second act resembles a speech about our expectations from the confronting candidates, about their strategies, their experience and behaviour in terms of speech-making (McKinney and Carlin 2004: 215). The final act brings about a speech concerning the debate that has just taken place, the candidate who performed the president part better, the winner of the debate, the intense moments, the blunders and errors that the candidates made (McKinney and Carlin 2004: 215).

Televised debate is not just the most intense dramatic moment of electoral communication. In the *hic et nunc* reality of the debate representations on situations, events and political actors are brought to focus. Practically, viewers are active observers of a *tableau vivant* of the political reality in society at some point. Is one of the reasons why the TV debate are "the most prominent shared campaign experiences of the voting public" (Boydstun et al. 2013: 254).

The televised debates remain essential forms of communication for the functioning of democracy (Beciu 2009: 139–140) and could be an ideal opportunity for candidates to establish an interpretation of reality, to impose control and symbolic representations of the situation in the political field.

12.3 Positioning Acts as Semiotic Practices in TV Debates

In an effort to define and build identity, any subject can take contradictory, inconsistent discursive positions: "[...] 'Identity', subjectivity, is thus a matrix of subject-positions, which may be inconsistent, or even in contradiction with one another" (Belsey 2002: 51). Michael Holquist, commenting Bakhtin's Dialogism, introduces the concept of "law of placement" in order to explain this phenomenon, which becomes relevant for understanding the identity-alterity relationship, I–Other:

> Bakhtin explicates this law with a just-so story that uses seeing as a means for grasping what is essentially a non-visual situation. He begins with a simple datum from experience; not an observer looking at trains, but an observer looking at another observer. You can see things behind my back that I cannot see, and I can see things behind your back that are denied to your vision. We are both doing essentially the same thing, but from different places: although we are in the same event, that event is different for each of us. Our places are different not only because our bodies occupy different positions in exterior, physical space, but also because we regard the world and each other from different centers in cognitive time/space. (Holquist 2002: 20)

According to Richard Holt "each person is in a unique place for seeing the world as the fundamental condition for understanding" (2003: 226). Any individual is located inside a system of positions that engage his identity and claim a correlative position (place) from the interlocutor. According to positioning theory (Davies and Harré 1990; Harré and Gillet 1994; Harré and van Langenhove 1998; Moghaddam et al. 2008; Moghaddam and Harré 2010; Harré and Moghaddam 2016), identity and self are discursive constructions. Positioning can be reproduced in social practices, bargained in various moments, discursive or ideological rules can be attached thereto, which make them socially recognisable:

> Institutions construct their ideological and discoursal subjects; they construct them in the sense that they impose ideological and discoursal constraints upon them as a condition for qualifying them to act as subjects. For instance, to become a teacher, one must master the discursive and ideological norms which the school attaches to that subject position - one must learn to talk like a teacher and "see things" (i.e. things such as learning and teaching) like a teacher. (Fairclough 1995: 39)

Initially, the concept of positioning has been introduced to replace the concept of *role* (from the dramaturgical perspective on communication), considered static and less flexible in understanding the discursive behaviour of social actors (Harré and Gillet 1994: 33–36; Harré and van Langenhove 1998: 14–17). Subsequently, authors stated that positioning theory must be understood in complementarity with the theory of role: "Positioning Theory is to be seen as complementary to the older framework of Role Theory, rather than an alternative approach" (Moghaddam et al. 2008: 9). Additionally, in certain situations, positioning acts can be "the birthplace of roles" (Moghaddam et al. 2008: 9). Therefore, the role can be regarded as the *continuous* aspect of the discursive performance of a social actor, and positioning is the *discreet* aspect of the discursive exchange. The balance between the continuous and the discreet aspect of communication is maintained by the very conflictual relationship between "seeing" and "understanding" (Marcus 2011: 183). It is the role that we rather observe. While we perceive the role's moments of discontinuity, of inadequacy, we better understand the conjectural discursive position (Drăgan 2016: 10).

To better understand the concept of positioning, we briefly discuss some other concepts close to its meaning. For example, in 1981, Goffman introduces the concept of *footing*, whose significance is associated to that of position and positioning (Davies and Harré 1990: 44; Moghaddam et al. 2008: 9). *Footing acts*, or interlocutors' roles (Beciu 2009: 43) are just an aspect of positioning. In Goffman's dramaturgical model, social actors are discursive characters, who perform *roles* through *footing mechanisms*. Another example comes from the tradition of the French school of discourse analysis. François Flahault introduces, in 1978, the concept of *places report* or *system of places* (*rapport de places* in French), which is close to the positioning concept significance-wise. For Flahault, establishing the relationships between places in a real discursive exchange occurs from a predetermined system of places, that is, from an *experiential matrix of places*. The speaking subject is inevitably caught within a (discursive) system of places. This system is recognisable in communication practice. Individuals are capable of recognising them, they have a certain discursive competence. It is on this relationship between places that the identity of speaking subjects is built. The illocutionary value of statements is the one that determines a certain configuration of the relationships between places in a discursive exchange. Words operate in a relational logic of the type "qui je suis pour toi, qui tu es pour moi" ("who I am for you, who you are for me") (Flahault 1978: 50). From this perspective, "Positioning theory is about how people use words (and discourse of all types) to locate themselves and others" (Moghaddam and Harré 2010: 2).

Built on the tradition of discursive construction of the self, the positioning theory can be described using a three-dimensional model. The three-dimensional model of positioning captures the relational aspects and the significance conditions of any dialogue interaction. The three (de)construction conditions of the significance of the discursive exchange determine each other; there is a dynamic relationship between them (Fig. 12.1).

The first aspect refers to positions and positioning. The way that various categories of people engage in discursive exchanges (*positioning acts*), the rights and

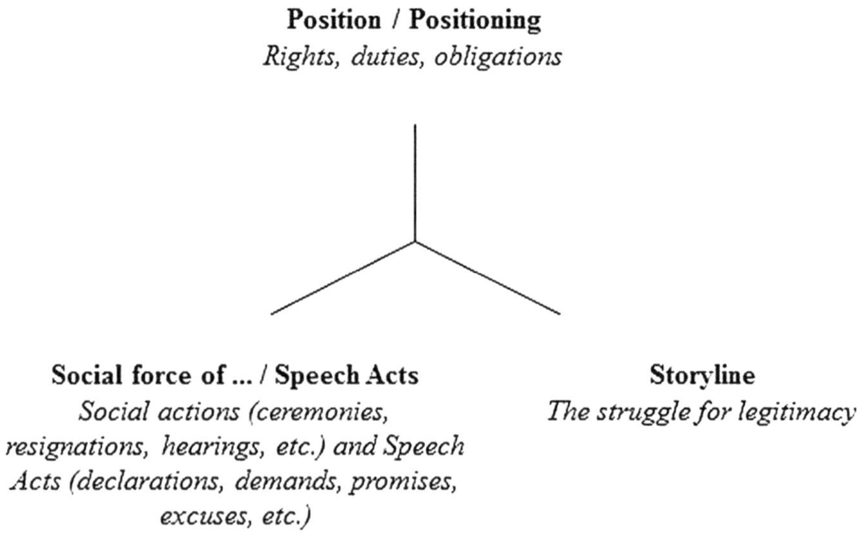

Fig. 12.1 The three-dimensional positioning model (Harré and van Langenhove 1998: 18; Moghaddam et al. 2008: 12; Moghaddam and Harré 2010: 50)

obligations they assume in their communication acts are "positions" (Harré and Gillet 1994: 34). They are predefined, by default. The rights, responsibilities and obligations associated to positions are set in the socio-discursive imaginary of the community, and they follow a certain moral order established within the society at a certain point: "Positioning Theory is the tool for exploring the relation between what is possible and what is permitted" (Moghaddam et al. 2008: 13). Moreover, positioning analysis allows us to capture how "rights and duties are distributed among the actors in the course of complex discursive interactions" (Moghaddam and Harré 2010: 6).

For example, when French presidential candidate Marine Le Pen portrayed her opponent as "out of touch and elitist" and "Mr. Macron is the candidate of savage globalisation" has not just made an offensive statement (attack) against the candidate Emmanuel Macron. She has put into practice a certain positioning strategy. Practically, as mentioned in the introduction, she indexed a certain type of social relationships built on the semiotic model of binary oppositions: elite/ordinary people, republican citizens/followers of globalization, etc.

It is important to note that "positions" are not socially regulated, they are not institutionalised. They are merely socially recognisable. Political actors have "the practical sense" (Bourdieu 1991: 179) of political game that allows them to recognise certain positions in the *experiential matrix of positioning*, or in the "semiotic web" of other discourses. This enables them to build positioning strategies in discursive interactions. Therefore, positioning can be regarded as a *semiotic configuration* with a certain stability in the communicational practices of the society.

The second aspect is about the way this positioning is activated. We introduce ourselves to the others, we build our discursive ego with individual narratives (*storylines*) that we employ in verbal interactions:

> Positioning can be understood as the discursive construction of personal stories that make a person's actions intelligible and relatively determinate as social acts and within which the members of the conversation have specific locations. (Harré and van Langenhove 1991: 395)

It is with these stories that we make our entrance in the context, we present ourselves to the others. An example in this regard may be a moment from the first 2016 presidential debate between Hillary Clinton and Donald Trump on September 26, when Donald Trump self-characterizes himself: "I have a winning temperament. I know how to win". Trump argues with a short story about how he built his empire: "[...] my father gave me a very small loan in 1975, and I built it into a company that's worth many, many billions of dollars" (Blake 2016). He activates in this speech sequence the narrative of the "Winner", of the "Hero".

The last aspect of a positioning act activates the concepts of "illocutionary force" and "speech acts". These are adopted from speech act theory, in order to describe the social significance of a discourse, gesture or social action (Moghaddam et al. 2008: 10). Any conversational exchange has an illocutionary value by default, a certain social strength. According to Rovența-Frumușani, the *institutional condition*, the position of power of the political actor in society, becomes an argument of legitimacy. The illocutionary force of political discourse, Austinian performativity shows itself in the effect of this position of power (Rovența-Frumușani 2012: 150–151). During the last presidential debate in Romania, Klaus Iohannis propose to his opponent, Victor Ponta, the following:

> Mr. Ponta, you are the president of PSD, I am the president of ACL. Let's make a business between us, men: Let's call the Parliament in session tomorrow, we can work the day after tomorrow, we remove the Amnesty and Pardon Law from Senate, we reject it and we green light all investigations and arrests. (O'Clock Nov. 13, 2014)

This is not a strategic statement with political content only, but it involves a strong ilocutionary force. According to Rom Harré and Grant Gillett, "Each speech-act has an illocutionary force, its social power as uttered in a certain context" (1994: 32). Therefore, positioning acts have the capacity to create social significance (Moghaddam and Harré 2010: 71).

Positioning theory reveals how the sensible space between observation and action can be used in an adequate, intelligent way (Drăgan 2016: 12).

The dynamic aspect, the internal metabolism that puts into motion various semiotic operations—the indexing of power relations, the mechanisms of interpretation, the potential of significance of the positions' configurations already established or experientially activated in the socio-discursive imaginary—the space of creativity occurring in the play of positional acts, allow the understanding of positioning practices as semiotic practices, as "meaning production processes" (Kristeva 1980: 251). From this perspective, positioning practices can be studied "not as an already established structure but as a *patterning*, as an apparatus that produces and transforms the meaning before it is already constituted and circulated" (Kristeva 1980: 252). Such an approach justifies the choice of multimodal analysis as a suitable methodological tool for the study of positioning practices.

In addition, the authors of positioning theory provide a method of analysing processes that occur between socially predefined convictions and practices and the individual way of acting in a particular situation. For example, positioning theory is an innovative tool that allows political actors to gain a better understanding and contribute in an appropriate manner "to conflict resolution through highlighting and assessing the story lines used in political processes" (Moghaddam and Harré 2010: 6). Basically, positioning theory describes the functional aspects of any discursive interaction in different communication contexts.

12.4 Research Methodology

12.4.1 The Positioning Analysis Model

According to the authors of positioning theory, a positioning act entails a two-step procedure. The first is *prepositioning*, an act that occurs by default when the interlocutors' attributes are known (or presumed). It consists in the moment of guessing "who you are talking to". We can speak of reflexive prepositioning (biographic elements, personal narratives, etc.) and prepositioning actions of the other (statements about the interlocutor's character, etc.). The next step is *positioning itself* in the dynamics of discursive exchange, and performing the assumed role.

In addition, the authors of positioning theory distinguish two levels of positioning, *first-order* and *second-order positioning*:

> First-order positioning refers to the way persons locate themselves (often implicitly) within an essentially moral order by using several categories and storylines, which is, in turn, predicated on an internalized discourse. In first-order positioning, the positioning act is perceived by a character (implicit) and a declaration (explicit). Second-order positioning (or *accountive positioning*) arises when an act is contested or negotiated in the course of the same conversation and occurs if first-order positioning is challenged or must be negotiated in some way. (Moghaddam and Harré 2010: 70–71; van Langenhove and Harré 1994)

When analysing positioning, certain research questions are at stake (Harré and Rossetti 2010: 114). Which are the narratives that each social actor brings into the context to legitimise himself? Which is the social significance of speech acts (illocutionary force), following the performance of individual narratives? Which rights and obligations have appeared in the discourse and how were they assigned, refused, accepted or contested by the speakers?

Therefore, the positioning analysis entails the following three steps (Harré and Rossetti 2010: 115–122):

1. Identifying prepositioning moves (reflexive prepositioning and the prepositioning of the Other).
2. Identifying and analysing individual narratives (e.g. the *Common person*—"I am one of you"; the *Saviour*, the *Hero*—"I have the solution, I am your saviour, follow me!"; the *Robin Hood syndrome*—"Moral justifications for certain political acts", etc.).
3. Confronting positional moves with the ideological profile of the social actors.

The analysis of prepositioning movements takes into account the dynamics of the debate. Since any debate can be analysed as a *drama in three acts,* I consider that it is useful to analyze the pre-positioning movements of political actors both in the important moments of the debate (such as "nodal" episodes of the TV debate), and during the first two acts preceding the debate. In the first situation, we are talking about the synchronic dimension or the discreet aspect of the prepositioning, and in the second situation about the diachronic dimension, or the continuous aspect of the prepositioning.

Practically, the first step of the positioning analysis consists in evaluating the communication situation, namely how "the target person has, or lacks, a desirable attribute for the ascription of a right or a duty relevant to the situation in hand" (Harré and Rossetti 2010: 115). From this point of view, for example, it is interesting to track the play of positions of the political and media actors during the presidential elections in Romania in November 2014, regarding the issue of the electoral debate. This is the metadebate-type discourse. For example, journalists have positioned themselves as "competitive actors on a TV market", paying less attention to the role of the media in "consolidating electoral debate as a practice in political and media culture" (Beciu 2015: 270). The candidates also engaged themselves in a "play of positions" that highlighted a process of "symbolic negotiation" (Beciu 2015: 276) of the status and the relationship in the political and media field.

The last step, which aims at confronting the positioning movements with the candidate's ideological profile, allows the leader's autonomy to be observed in relation to the political family that supports him, the degree of freedom to the reference.

The voters themselves are involved in the positioning play. For example, in the case of a negative vote—which is rather oriented against somebody that in favour of a candidate—the authors of the positioning theory suggest that "we have a nice case of meta-positioning—positioning oneself as having a higher duty than the first-order duty of supporting the candidate one believes would be the best in office" (Moghaddam and Harré 2010: 5). Therefore, the mechanism itself of the voting decision can be regarded as a play of positions that allows the conduct of a comparative act between the candidates. In an environment where the trust in parties decreased (Benoit 2014: 10) it is important to understand the vote as a comparative choice among candidates. The voting decision is rather influenced by the individual candidates and their teams, the way they are perceived by the public. Loyalty to a party, though still important, has a lesser influence on the voting decision (Benoit 2014: 10).

By analysing positioning, the dynamic relationship between the three dimensions of the positioning of social actors during a discursive exchange (individual narratives, social significance and position) are captured. The proper positioning, along with an authentic performance (credible individual narratives) make the social significance of the candidates' discursive behaviour acquire consistency in the audience's view.

12.4.2 The Multimodal Analysis Model

For the second level of research design I propose a multimodal analysis model. An interdisciplinary approach—analysing the positioning complementarily with a professional framework for multimodality research, ELAN—allowed on the one hand, capturing the dynamics of the discursive exchange in the considered episode, and on the other hand, to disambiguate the positioning (and the interpretation thereof) of the political actors involved in the debate. This is the foundation on which I perform positioning analysis in the present chapter.

In this study, we took into consideration three extracts from a video recording of the last presidential debate in Romania, broadcasted by B1TV on the 12th of November 2014. The first two sequences—lasting one minute and 30 s respectively

—capture an exchange of replies, from the middle of the debate, on one of the most important subjects of the political-media agenda, namely the theme of "Justice". The last sequence, lasting one minute and 30 s, at the end of the debate, is an exchange of replicas between the candidates, of the *ad hominem* attacks type. The protagonists were Victor Ponta, incumbent Prime Minister at the time and SDP (The Social Democratic Party) president, and Klaus Iohannis, former mayor of Sibiu, PNL (The National Liberal Party) president. During the final televised debates for presidential elections in Romania in recent years, such key moments (sequences) that decided the winner's fate have been frequent. We proceeded on the assumption that such episodes concentrate the candidates' discursive strategies, and the manner in which protagonists manage these moments can be decisive in directing the viewers' preferences. Selected speech sequences highlight, as mentioned in section three, the two modalities of positioning political actors in discourse: a position towards the content of communication (what it enunciates)—for example on a common theme, "Justice"—and a (relative) position towards the participants to the discursive act.

Multimodal analysis allows understanding how inter-semiotic relationships (or inter-modal) are established between different *semiotic modes*. Understanding how different semiotic modes work together is the central objective in any multimodal research (O'Halloran 2011: 121; Siefkes 2015: 114). In multimodal texts, meanings conveyed by verbal language integrate the meanings of the different semiotic modes for the construction of meaning. Kress and van Leeuwen (2010) use the term "mode", while O'Halloran (2011) prefers the term "semiotic resource". In our multimodal analysis model, we have used three categories of semiotic modes: verbal language, gesture and cinematic resources (cinematic frames and horizontal perspective, or camera angle).

The multimodal analysis model that we have adopted involves the following steps:

(1) The annotation of verbal resources, gesture resources and cinematic resources using ELAN.
(2) Identifying semiotic types of gestures.
(3) Attributing function to gesture (Colletta et al. 2009: 61–62).
(4) Identifying the relationship of gestures with corresponding verbal discourse (semantic relations).
(5) Identifying the semiotic modes that political actors use to build their individual narratives and discursive position.
(6) Comparative analysis of movements and positioning strategies.

The data were annotated and analyzed using ELAN, a multimedia annotation tool developed at the Max Planck Institute for Psycholinguistics in Nijmegen, The Netherlands.

To study the gestures, The MUMIN coding scheme was used as the starting point. This scheme was developed in the Nordic Network on Multimodal Interfaces (Allwood et al. 2007). The MUMIN coding scheme is a general framework for the study of gestures in interpersonal communication. Practically, such a methodological framework allows analysis of the dynamics of discursive exchange among interlocutors: "The MUMIN scheme was developed to study the communicative functions of feedback, turn management and sequencing" (Navarretta and Paggio 2013: 104). For example, Navarretta and Paggio (2013) have used this methodological framework to analyze several sequences during two TV debates: the first is from the third debate in the 2008 American presidential campaign and the second is from the first prime minister debate in the 2010 British election campaign.

In addition to the MUMIN coding scheme, for the annotation of hand gestures, I used a coding system which is generally used in the analysis of various forms of political communication (Maricchiolo et al. 2012: 408; 2013:117; Gnisci et al. 2013: 881), presented in Table 12.1:

In this taxonomy of hand gestures, hierarchically organized, "The term 'link' here is intended not in terms of gesture-speech 'co-occurrence', but with reference either to the 'content' of the speech or to the 'structure' of the speech" (Maricchiolo et al. 2012: 407).

In order to identify the "relational meanings" (Rovenţa-Frumuşani 1999: 193) that appear in the interactions between the semiotic modes, various research models can be used. We used in this study the taxonomy of semantic relations proposed by Colletta et al. (2009: 62–63) in order to establish the correlations between the gestures and the corresponding verbal discourse:

- Reinforces, in which the informational content of the gestures is identical and confirms the information of the verbal message;
- Complements, by which the gestures disambiguate the verbal message, or "cause a nuance of attitude in relation to the facts evoked by verbal discourse" (Rovenţa-Frumuşani 1999: 193–194);
- Supplements, through which gestures supplement the information conveyed by the verbal message;

Table 12.1 Taxonomy of hand gestures (Maricchiolo et al. 2012: 408; Gnisci et al. 2013: 881)

Type	Macro-categories (or functions)	Specific-categories (or attribute)	Sub-categories (or semiotic value)
Speech linked gestures (SLG)	Conversational (interactional)	Cohesive	
		Rhythmic	
	Ideational	Emblems	
		Illustrative	Deictic
			Metaphoric
			Iconic
Speech nonlinked gestures (SNG)	Adaptors	Self-adaptors	
		Object-adaptors	

- Integrates, through which the meanings of the verbal message become more accurate due to the imagistic properties of the gestures;
- Contradicts, in which the information conveyed by the semiotic modes is contradictory;
- Substitutes, in which "the information provided by the gesture replaces linguistic information, as in the case of certain performative and interactive gestures" (Colletta et al. 2009: 63).

For each of the situations subject to analysis, the gesture-word correlation has been disambiguated in the context by comparing the informative content of the verbal discourse to that conveyed by the gestures performed by that political actor. Information conveyed by a particular gesture correlated with the corresponding verbal message can be framed in one or more types of semantic relationships. Gestures can carry out more relational functions, and they implicitly have more relational significance depending on the communication context (Allwood et al. 2007: 279). In such situations, we selected the most visible semantic relationship (out of the six), accompanied by some comments.

12.4.3 Results and Discussion

Figure 12.2 shows a screenshot of the ELAN user interface, corresponding to the first selected speech sequence. In this sequence, we have an example of positioning of candidate Klaus Iohannis towards the content of communication (what it

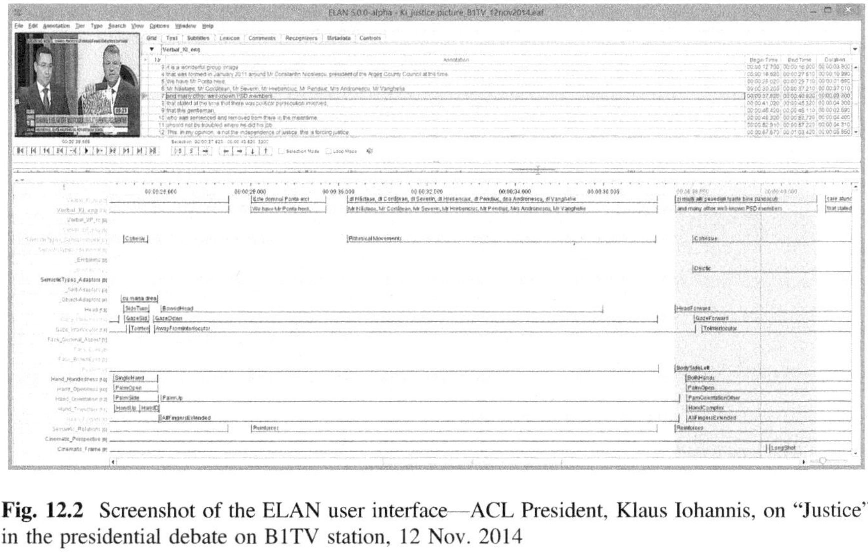

Fig. 12.2 Screenshot of the ELAN user interface—ACL President, Klaus Iohannis, on "Justice" in the presidential debate on B1TV station, 12 Nov. 2014

enunciates)—on the common theme, "Justice"—and a (relative) position towards his opponent, SDP leader Victor Ponta.

At the bottom of the screen, on the left, we can see the entire coding scheme used in this study, with the three categories of semiotic modes: verbal language, gesture and cinematic resources (cinematic frames and horizontal perspective). The verbal level of our annotation scheme includes only the orthographic transcription of the discursive exchange in the considered sequence, in Romanian, and the English translation, for each of the political actors. In this study I focus on the identification of the semiotic modes that political actors use to build their individual narratives and discursive positionings. Below I also present a comparative analysis of movements and positioning strategies of political actors in the discourse sequences considered.

In this episode, candidate Klaus Iohannis attempts to exploit one of his adversary's vulnerabilities regarding corruption in the party. The liberal leader presents a photo from January 2011, in which Victor Ponta appears alongside important Social Democratic leaders at a demonstration supporting the former president of Argeş County Council, Constantin Nicolescu, convicted by the Romanian Judiciary for acts of corruption.

The President of the Christian Liberal Alliance (ACL), Klaus Iohannis, activates the "Knight of Justice" narrative in this episode, which reveals the acts and deeds of corruption of the "Power" to the public. The meanings of the verbal message are supplemented with the meaning potential of the deictic pointing gesture that Iohannis makes in front of the cameras. Practically, the information transmitted by the iconic object (photograph) is added to the information of the verbal message. Thus, the act of positioning gains strength and relevance for the public. In this episode, Klaus Iohannis updates certain representations that the public can easily access through the iconic object. The ACL leader selects from the content plane (the photographic content) those pertinent features (the candidate Victor Ponta by Social Democratic members convicted for corruption) that allow him to index certain power relations already established in the collective imagination. In this situation, it is about the image and positioning of the Social Democratic "barons" regarding the theme of "Justice".

Figure 12.3 shows a screen capture from the sequence immediately following the earlier episode.

In this sequence I aim to explicate the answer and the correlative positioning of candidate Victor Ponta regarding the same theme, "Justice". The Social Democratic leader read in front of the cameras, without keeping a visual contact with his adversary or viewers, Article 133 of the Constitution of Romania governing the independence of the Judiciary. In a comparative assessment, it is easy to find an answer to the questions: Which of the two acts of communication is more relevant in meaning and richer in higher illocutionary force? Is it the first act, in which the information of the verbal message is supplemented with the information of the deictic pointing gesture or the second, representing the act of reading a text in front of the audience present only at the virtual level in the enunciation scene?

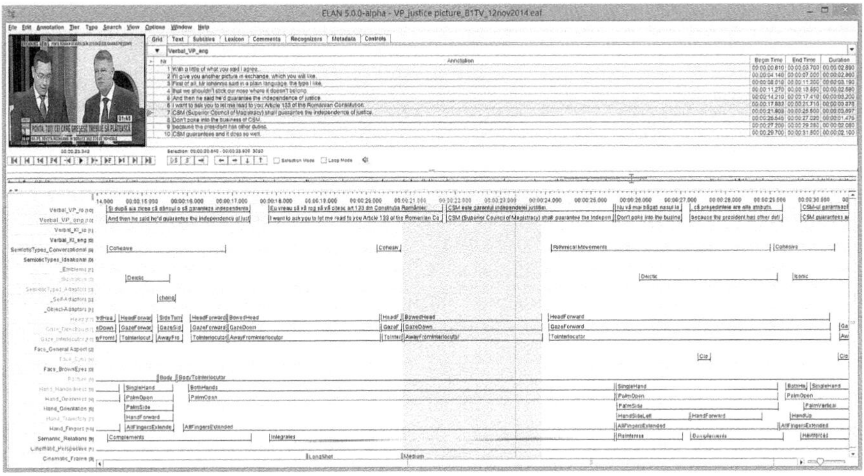

Fig. 12.3 Screenshot of the ELAN user interface—SDP leader Victor Ponta's response to the common theme of "Justice" in the presidential debate on B1TV station, 12 Nov. 2014

Figure 12.4 shows a screenshot of the ELAN user interface, corresponding to the last selected speech sequence at the end of the debate.

The screen capture shown in Fig. 12.4 captures one of the moments when Victor Ponta's candidate builds the "Expert" narrative. It is one of the favorite narratives of the incumbent presidential candidate during the two presidential debates. Moreover, from the analysis of the speech acts of Victor Ponta, combined with the significance of gestures provided by the multimodal analysis—for instance, the piece of paper

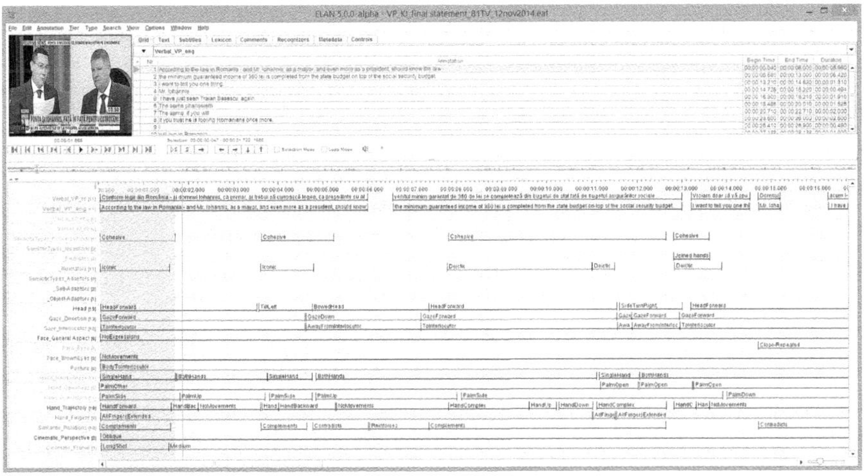

Fig. 12.4 Screenshot of the ELAN user interface—final statements, B1TV station, 12 Nov. 2014

he shows to his interlocutor—in this discursive sequence, he takes on the position of *expert*, with the related rights and obligations (competent, responsible, knowledgeable, capable of coherent action) *versus* a pre-positioning of the other, his opponent Klaus Iohannis, as a *dilettante, pupil* or *trainee* (not knowledgeable).

In the sequences immediately following that shown in Fig. 12.4, Victor Ponta suddenly changes his narrative, activating the narrative of "Hero" speaking on behalf of the "fooled" people. *The Hero* identifies the "public enemy", Traian Băsescu, the former Romanian president, in the person of his interlocutor Klaus Iohannis. This association of images was basically the main strategy that the candidate of the governing party adopted throughout the election campaign:

> I want to tell you one thing. Mr. Iohannis, I have just seen Traian Băsescu, again. The same phariseism, if you trust he is fooling Romanians once more. I will live in Romania without Băsescu and Iohannis. (O'Clock Nov. 13, 2014)

The selected speech sequences are relevant to the candidate's discursive behavior during the two presidential debates. The positioning acts built by Klaus Iohannis show some stability around the "Knight of Justice" narrative, whether we are talking about the theme of "Justice", "Corruption", or the theme of "Diaspora Voting", which are practically the dominant themes of the debates. Moreover, the ACL candidate uses much less semiotic resources, but in a more coherent way than his adversary. This can be interpreted as a rather "sober" performance, inherent of a better knowledge of the presidential role. In addition, the audience has access to a more "airy" reading space and may feel a certain coherence and authenticity in the positioning acts of candidate Klaus Iohannis.

Instead, Victor Ponta's positional play regarding the topics in question appears more contradictory, ambiguous and harder to decipher in relation to his adversary's positional strategy. For example, in earlier analysed sequences, Victor Ponta proposes to the public the narrative of the "Expert" complementary to the "Hero-Savior" narrative in the same communication situation. Such a construction can create ambiguity in the interpretation scenarios of the audience.

12.5 Conclusion

As explained in the previous section, the practice of integrating the meanings of different semiotic modes allows for a better understanding of verbal language, actions, positioning acts, and behavioral type of political actors involved in discursive interactions.

During the two televised debates the candidate of the incumbent party mostly wanted *to seem that he was a certain way*, often omitting the aspect of authentic involvement in the dialogue. The candidate of the opposition's party *looked as he was*, relying on emphasising touchy, conjectural topics, to the detriment of others.

The journalist Cristian Tudor Popescu[3] captured this aspect in an article published in the daily newspaper Gândul (*The Thought*), on the day of the second televised debate:

> The presence of Mr. Iohannis in the show last night on RealitateaTV is contradicted blatantly by his own image, the one he struggles to deliver to voters for months: the settled, calm, serious "German", the man of "things well done". [...] Victor Ponta wanted to pass for a teacher asking a student loads of questions. He threw a wave of numbers and statements over Iohannis, in some cases adding and bluffing, as the verifications of the newspaper Gândul show. And the mimic of his face, his gestures, his fussing, his chattering, his eye movements made him look like a fretting apprentice. Mr. Iohannis looked like a teacher, even like a master: he spoke much less, with long breaks between the words spoken, *magister dixit*, showing an impenetrable face and a fixed gaze. On the contrary, the content of his utterings was that of a scoundrel student, who is trying to get away with cheap verbal tricks. In short, Ponta was the way we knew him, bold, superficial, cunning, boarish, Iohannis was unlike what we knew of him – improvised, inconsistent, bombastic, the man of things done in a rush. Victor Ponta did not manage to save his appearance, Klaus Iohannis revealed, for the first time, his essence" (November 12, 2014)

The positioning analysis captures the dynamic relationship between individual narratives, the social significance and position of the political actors in a discursive exchange. Complementarily, the multimodal analysis, presented for illustrative purposes in this article, facilitates noticing how social actors make use of various semiotic resources to assign themselves certain positions during verbal interaction, which the correlative positions are, as well as the manner in which the two candidates build their discursive ego. In short, it is useful for deconstructing the discursive behaviour of the social actors in key moments of the televised debate. This was possible using an interdisciplinary approach: analysing the positioning complementarily with the ELAN professional framework for multimodality research. This methodological framework allowed, on the one hand, the capturing of the dynamics of the discursive exchange in the episode considered, and on the other hand, to disambiguate the positioning (and the interpretation thereof) of the political actors involved in the debate.

References

Allwood, Jens, Loredana Cerrato, Kristiina Jokinen, Costanza Navarretta, and Patrizia Paggio. 2007. The MUMIN coding scheme for the annotation of feedback, turn management and sequencing phenomena. *Language Resources and Evaluation* 41 (3/4): 273–287.

BBC News. 2017. French election gets dirty: Insults that marked a fierce debate. *BBC News*, 4 May 2017. http://www.bbc.com/news/world-europe-39803333/. Accessed 7 May 2017.

Beciu, Camelia. 2009. *Comunicare şi discurs mediatic. O lectură sociologică* [*Communication and media discourse. A sociological reading*]. Bucureşti: Comunicare.ro.

[3]Cristian Tudor Popescu is a Romanian journalist, essayist and short-story writer. Four years in a row, in 2005–2008, he was designated the best Romanian journalist.

Beciu, Camelia. 2015. Dezbaterile Electorale şi Rolul Mediei în Campania Prezidenţială 2014 din România [Electoral debates and the role of the media in the presidential campaign 2014 in Romania]. *Revista Română De Sociologie [Romanian Journal of Sociology]* 26 (3): 253–278.
Belsey, Catherine. 2002 [1980]. *Critical practice*, ed. by Terence Hawkes, 2nd edn. London: Routledge.
Benoit, L.William. 2014. *Political election debates: Informing voters about policy and character*. Lanham, U.K.: Lexington Books.
Blake, Aaron. 2016. The first Trump-Clinton presidential debate transcript, annotated. *The Washington Post*, September 26, 2016. https://www.washingtonpost.com/news/the-fix/wp/2016/09/26/the-first-trump-clinton-presidential-debate-transcript-annotated/?utm_term=.da4fa58cf170/. Accessed 7 May 2017.
Bourdieu, Pierre. 1998. *On television*. New York: New Press.
Bourdieu, Pierre. 1991. *Language and symbolic power*. Cambridge: Polity Press.
Boydstun, E. Amber, Glazier, A. Rebecca, and Matthew T. Pietryka. 2013. Playing to the crowd: Agenda control in presidential debates. *Political Communication* 30 (2): 254–277.
Colletta, Jean-Marc, Kunene, N. Ramona, Venouil, Aurélie, Kaufmann, Virginie, and Jean-Pascal Simon. 2009. Multi-track annotation of child language and gestures. In *Multimodal Corpora. From models of natural interaction to systems and applications*, eds. Michael Kipp, Jean-Claude Martin, Patrizia Paggio and Dirk Heylen. 54–72. Berlin: Springer.
Davies, Bronwyn, and Rom H. Harré. 1990. Positioning: The discursive production of selves. *Journal for the Theory of Social Behavior* 20 (1): 43–63.
Drăgan, Nicolae-Sorin. 2016. Loc, rol şi poziţie. Practici discursive în discursul politic românesc [Place, role and positon. Discursive practices in Romanian political discourse]. *Romanian Review of Young Researchers (RRYR)* 4 (I): 7–20.
Eco, Umberto. 1976. *A theory of semiotics*. Bloomington: Indiana University Press.
Erickson, V. Keith. 1998. Presidential spectacles: Political illusionism and the rhetoric of travel. *Communication Monographs* 65 (2):141–153.
Erickson, V. Keith. 2000. Presidential rhetoric's visual turn: Performance fragments and the politics of illusionism. *Communication Monographs* 67 (2): 138–157.
Fairclough, Norman. 1995. *Critical discourse analysis: The critical study of language*. London: Longman.
Flahault, François. 1978. *La parole intermédiaire [La Parole intermediate]*. Paris: Seuil.
Gnisci, Augusto; Maricchiolo, Fridanna, and Marino Bonaiuto. 2013. Reliability and validity of coding systems for bodily forms of communication, In *Body—Language—Communication: An international handbook on multimodality in human interaction, Handbooks of Linguistics and Communication Science 38.1.*, Volume 1, eds. Cornelia Müller, Alan Cienki, Ellen Fricke, Silva H. Ladewig, David McNeill and Jana Bressem. 689–706. Berlin: De Gruyter Mouton.
Goffman, Erving. 1981. *Forms of talk*. Philadelphia: University of Pennsylvania Press.
Harré, H. Rom, and Luk van Langenhove. 1991. Varieties of positioning. *Journal for the Theory of Social Behaviour* 21 (4): 393–407.
Harré, H. Rom, and Grant Gillet. 1994. *The discursive mind*. London: Sage.
Harré, H. Rom, and Luk van Langenhove (eds.). 1998. *Positioning theory: Moral contexts of intentional action*. Oxford: Blackwell Publishing.
Harré, H. Rom, and Mark Rossetti. 2010. In *words of conflict words of war. How the language we use in political processes sparks fighting*, eds. Fathali Moghaddam and Rom H. Harré. 111–123. Santa Barbara, CA: Praeger.
Harré, H. Rom, and Fathali M. Moghaddam (eds.). 2016. *Questioning causality: scientific explorations of cause and consequence across social contexts*. Santa Barbara, CA: Praeger.
Hellweg, A. Susan; Pfau, Michael, and Steven R. Brydon. 1992. *Televised presidential debates: Advocacy in contemporary America*. New York: Praeger Publishers.
Holquist, Michael. 2002 [1990]. *Dialogism: Bakhtin and his world*. London: Routledge.
Holt, Richard. 2003. Bakhtin's dimensions of language and the analysis of conversation. *Communication Quarterly* 51 (2): 225–245.

Kress, Gunther, and Theo van Leeuwen. 2010 [1996]. *Reading images: The grammar of visual design*. London: Routledge.

Kristeva, Julia. 1980. Problemele structurării textului [Problèmes de la structuration du texte]. In *Pentru o teorie a textului, Antologie Tel Quel* 1960–1971 [Trans. from volume *Tel Quel: Théorie d'ensemble*, Paris: Editions du Seuil, 1968, 297–316]. 250–272. Bucureşti: Univers.

Lempert, Michael, and Michael Silverstein. 2012. *Creatures of politics: Media, message, and the American Presidency*. Bloomington: Indiana University Press.

Marcus, Solomon. 2011. *Paradigme Universale* [*Universal Paradigms*]. Bucureşti: Paralela 45 Publishing House.

Maricchiolo, Fridanna, Augusto Gnisci, and Marino Bonaiuto. 2012. Coding hand gestures: A reliable taxonomy and a multi-media support. In *Cognitive behavioural systems 2011, LNCS 7403*, ed. Anna Esposito, Antonietta M. Esposito, Alessandro Vinciarelli, Rüdiger Hoffmann, and Vincent C. Müller, 405–416. Berlin, Heidelberg: Springer.

McKinney, S. Mitchell, and Lamoureux, E. R. 1999. Citizen response to the 1996 presidential debates: Focusing on the focus groups. In *The electronic election: Perspectives on the 1996 campaign communication*, eds. Kaid, Lynda Lee, and Dianne G. Bystrom. 163–177. Mahwah, NJ: Lawrence Erlbaum.

McKinney, S. Mitchell, and Diana B. Carlin. 2004. Political campaign debates. In *Handbook of political communication research*, ed. Kaid, Lynda Lee. 203–234. New Jersey (USA): Lawrence Erlbaum Associates, Inc. Publishers.

Moghaddam, M. Fathali; Harré, H. Rom, and Naomi Lee (eds.). 2008. *Global conflict resolution through positioning analysis*. New York: Springer.

Moghaddam, M. Fathali, and Rom H. Harré (eds.). 2010. *Words of conflict words of war*. Santa Barbara, CA: Praeger.

Navarretta, Constanza, and Patrizia Paggio. 2013. Multimodal Behaviour and Interlocutor Identification in Political Debates. In *Multimodal communication in political speech. Shaping minds and social action, international workshop, political speech* 2010, eds. Isabella Poggi; Francesca D'Errico; Laura Vincze, and Alessandro Vinciarelli. 99–113. Berlin, Heidelberg: Springer.

Nine O'Clock. 2014. Ponta-Iohannis televised duel continued on B1TV Wednesday night. *Nine O'Clock*, November 13, 2014. http://www.nineoclock.ro/ponta-iohannis-televised-duel-continued-on-b1tv-wednesday-night/. Accesed 10 Apr 2017.

O'Halloran, L. Kay. 2011. Multimodal discourse analysis. In *Companion to discourse analysis*, ed. Ken Hyland and Brian Paltridge, 120–137. London: Continuum.

Pfau, Michael. 2002. The subtle nature of presidential debate influence. *Argumentation & Advocacy* 38: 251–261.

Pleşu, Andrei. 2010. Parabolele lui Iisus. O analitică a receptivităţii [The parables of Jesus. An analytics of receptivity]. *Conference given on 27 October 2010, in the Aula of the "Alexandru Ioan Cuza" University of Iaşi*. Available at https://vimeo.com/17047701/. Accessed 27 Dec 2015.

Pleşu, Andrei. 2015. Interviewed by Daniela Zeca-Buzura, Micdejun cu un campion [Breakfast with a champion]. *TVR2*, May 9, 2015. http://www.tvrplus.ro/editie-mic-dejun-cu-un-campion-332455/. Accessed 20 Jan 2016.

Popescu, Cristian Tudor. 2014. Magistrul Loază şi ucenicul Forfotă [Master Scoundrel and apprentice Fuss]. *Gândul*, November 12, 2014. http://www.gandul.info/puterea-gandului/magistrul-loaza-si-ucenicul-forfota-13537738/. Accessed 13 Nov 2014.

Postman, Neil. 1985. *Amusing ourselves to death: Public discourse in the age of show business*. New York: Penguin.

Rovenţa-Frumuşani, Daniela. 1999. *Semiotică, Societate, Cultură [Semiotics, Society, Culture]*. Iaşi: Editura Institutului European.

Rovenţa-Frumuşani, Daniela. 2012. *Analiza discursului. Ipoteze si ipostaze* [*Discourse Analysis. Hypotheses and hypostases*]. Bucureşti: Editura Tritonic.

Siefkes, Martin. 2015. How semiotic modes work together in multimodal texts: Defining and representing intermodal relations. *10plus1: Living Linguistics* 1: 113–131.

Chapter 13
Racism and Classism in Mexican Advertising

An Exhibition of Visual Messaging

Carl W. Jones

Abstract This paper explores how Mexico's population has been faced with the polemics of class and race. This division continues today through the Mexican ruling class's appropriation of advertising. I am interested in the functions and systems in place that allow this to propagate and how meaning is being reproduced unperceived by the audience. My thesis question asks, *What are the visual representations of the power relationships in Mexico's political economy as reflected through the appropriation of advertising?* To answer this question, I perform a semiotic analysis of branded advertising messages created by the companies Bimbo, Palacio de Hierro and FEMSA, owned by the Mexican ruling families Servitje, Baillares and Garza respectively. Each television commercial is examined for signs, cultural codes, gestures, gaze and word tracks. These signs are decoded, and the conclusion is expressed through "An Exhibition of Visual Messaging", designed to inform the Mexican public of how messages are constructed and received, empowering the viewer to interpret and challenge the meaning behind the communications they are receiving through the metamedia.

13.1 Introduction

Classist and racist discourse is occurring in various parts of the world and particularly in Mexico. I argue that strategies that reinforce segregation, classist, and racist ideologies exist within Mexico's society, and these themes continue today through various forms, specifically media. I am interested in the functions and systems in place that allow the classist and racist discourse to propagate and how meaning is being reproduced unperceived by the audience. This project and its

C. W. Jones (✉)
Ryerson University, Toronto, ON, Canada
e-mail: carlwj1@mac.com

C. W. Jones
York University, Toronto, ON, Canada

A. Olteanu et al. (eds.), *Meanings & Co.*, Numanities - Arts and Humanities in Progress 6, https://doi.org/10.1007/978-3-319-91986-7_13

research sheds new light on the profession of advertising practitioners and scholars because advertising is often looked upon negatively in communication studies. Arguably, there is a broad spectrum of literature, such as Marx and Engels ("The Ruling Class and the Ruling Ideas" 2006: 9), Fuchs ("Dallas Smythe Today—The Audience commodity…" 2005) and the Frankfurt School's Adorno and Horkheimer ("Dialectic of Enlightenment" 2002: 94–136), that would regard advertising as a weapon of corporations, used to convince the public to consume commodities, and also to become an "audience commodity", as Smythe states (1977). However, I would like to demonstrate that advertising tools and techniques can be employed for the human good by creating messages that counter Western corporate communications and encourage local culture. This will be achieved through publishing this thesis project and informing the Mexican public of how messages are constructed and received, thereby empowering the viewer to interpret and challenge the meaning behind the communications that they are receiving through the metamedia.

Another contribution will be to help promote the study of advertising theory in Canada and Mexico. This project will strengthen the mutual understanding between these two approaches by using my own experience as a practising art director and scholar in Canada and Mexico. A final contribution will include approaching the mass media to publish my findings and holding an exhibition directed towards the general public, advertising practitioners and academics in communication studies.

Through performing semiotic research and expressing the results in a contemporary visual representation, my thesis project explores the *visual representations of the power relationships in Mexico's political economy as expressed in advertising* and how these relationships can be resisted.

Theoretical component

Five hundred years ago, the ruling class in Mexico categorized its inhabitants into racial categories, with 'indigenous' being the lowest and 'Spanish-born' the highest. This was reflected in the *pinturas de castas* that I will be discussing later on in this chapter. After the Spanish colonization, the races began to mix, creating a new Mexican-born group named *Mestizos*. These categorizations of Mexican citizens are reflected in the advertising messages produced inside Mexico's national boundaries and are financed by the ruling class. *Semiotic theory* and *critical race theory* applied to my research focuses "[…] on the discourse and images that create, maintain, and transform cultural relations of meaning and power as racial phenomena" (Lindlof 2011: 63).

During my research into the subject of race in Mexico, I discovered that racial categories have been an important theme in Mexican history. However, since 1907, the government has not officially classified its citizens on the basis of race; instead, it asks Mexicans to self-identify (Lizcano Fernández 2005) through language such as indigenous dialects.

The idea that white skin was an imported concept in Mexico that arrived through the colonialism of the Spanish is not true, as "[…] the preference for lighter skin existed worldwide before colonization amongst the Aztecs and in early Arab cultures too, as symbols of higher social class and as symbols of youth and virginity" (Hussain 2010: 19).

One widely-believed myth regarding a white god is attributed to when the Aztec emperor Montezuma welcomed the Spanish conqueror Cortez into Tenochtitlan, now Mexico City. The Aztecs believed that the god Quetzalcoatl would return as a pale-skinned bearded god-king, and Montezuma believed that Cortez was Quetzalcoatl, riding "[...] on the backs of animals resembling deer" (León-Portilla 1992: 3–11). Not all scholars agree with the Aztec legends about white gods, speculating that they were myths created by the Franciscan Monks after the conquest to help the Aztecs understand why they were defeated (Townsend 2003: 659–687).

Cortez had a slave named Malinche, who could speak Mayan and Nauhatal, the language of the Aztecs. She translated for him, and they became lovers. Malinche is known as the mother of modern Mexico because of her union with Cortez, which signified the mixing of European and Indigenous bloods (Fig. 13.1). This union created the racial and cultural category of 'Mestizo', which represents 83% of the Mexican population (CIA 2015).

Fig. 13.1 Cortes y la Malinche (1926) Painter: José Clemente Orozco

During the early 17th century, after the Spanish had the local population under their control, they began to classify the citizens into "castas" (categories), a hierarchical system of racial classification. It placed the 'Spanish-born' at the top, giving them the most rights, and the indigenous at the bottom. "Castas" were "[…] the product of the mixing of Spanish with Crioles, Indigenous, Mestizos and Blacks" (INAH 2009). In the 18th century, this classification was represented through paintings: "las pinturas de castas" (paintings of castes). Each painting depicts a couple and their offspring, demonstrating skin colour, clothing and customs. These paintings also represent "[…] the pyramid structure based on race, where the Spanish-born had all the political power" (INAH 2009). These constructed images favoured the native-born Spanish, as demonstrated in Fig. 13.2, where the figures are presented in a dignified manner, which is very different than the image of the mixed-race couple depicting an "Indigenous" male fighting a "mulatta" who has a child tied to her back (Fig. 13.3).

In the early 1800s, Charles Darwin travelled to South America and encountered the indigenous in Tierra del Fuego, describing them as "[…] the most abject and miserable creatures he had ever encountered […]" (Carrillo Trueba 2009: 27). This same "[…] repulsion at the indigenous peoples was echoed by the elite Mexicans who were considered white" (Carrillo Trueba 2009: 27). At the end of the

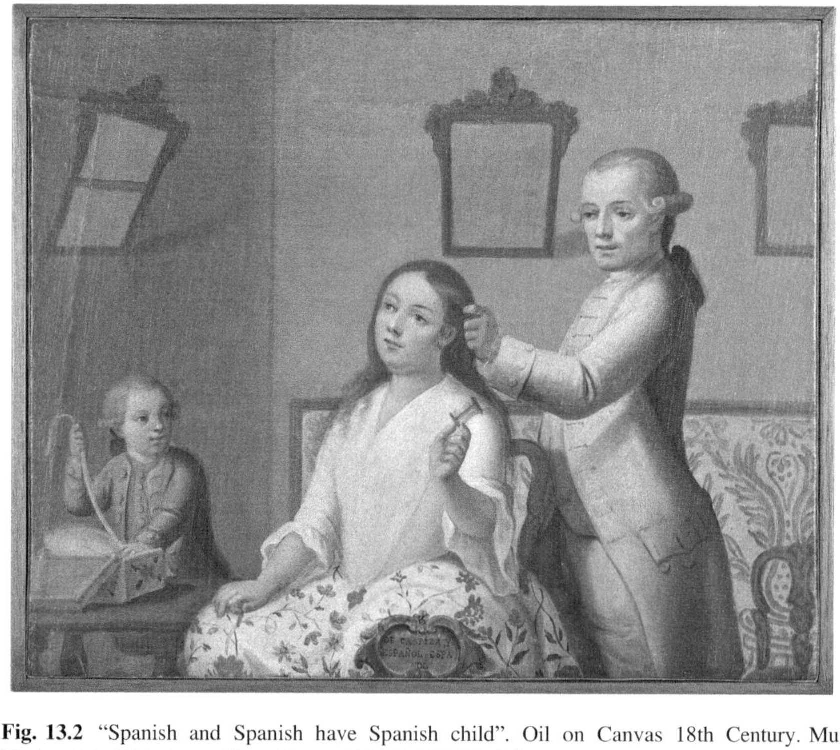

Fig. 13.2 "Spanish and Spanish have Spanish child". Oil on Canvas 18th Century. Museo Nacional de Historia en Chapultepec. (INAH. RF-341-15)

FOTOGRAFIA: GERARDO CORDERO A.

Fig. 13.3 "Indigenous and Mulatta have a throwback child" Oil on canvas. 18th Century. Museo Nacional de Historia en Chapultepec. (INAH. RF-341-15)

19th century, the concept of improving the human race through genetics, based on the theory of evolution, emerged; it was called 'eugenics'. In 1931, the *Ateneo de Ciencias y Artes, y la Sociedad Mexicana de Eugenesia para el Mejoramiento de la Raza* (Academy of Sciences, and Arts and the Mexican Society of Eugenics for the Betterment of the Race) was created (Carrillo Trueba 2009: 27); it was "[...] made up of doctors that discussed and created measures to improve the population" (Carrillo Trueba 2009: 27). One of its main proposals was to whiten the Mexican population. This was to be done through the "[...] immigration of white people and not Chinese or Blacks" (Carrillo Trueba 2009: 48–49) and by convincing the new immigrants to move to indigenous areas of Mexico (Carrillo Trueba 2009: 42–44).

The concept of lighter skin superiority continues to the present day, where, I argue, that in popular culture that the "social segregation" of the races and classes is reflected on national television. The majority of actors have a '*tez clara*' (clear complexion), apart from some typical characters, such as the maid in soap operas and actors in some government ads aimed at the general population (Carrillo Trueba 2009: 58). A government study by CONAPRED (National Coalition to Prevent Discrimination) was published in the Mexico City newspaper MIRA! on August 3, 2015, with a headline stating, "Mexicans are closet racists" (MIRA). Featured in the article is an interview with Sonia Rios Freije, Adjunct Director of CONAPRED, that was accompanied by various facts from their latest study: “4 out of 10 Mexicans felt they were treated differently because of their skin colour; [...] 55% admit that in Mexico you can be insulted for the colour of your skin; [...] and 20% are not satisfied with their own skin colour” (CONAPRED 2010; MIRA 2015).

A person, however, can have darker skin and be considered high class in Mexico. Having a clear complexion does not automatically give a person “class.” There is a popular saying, “*guerro del rancho*”, which means a “white person from the country”, or in other words, a white person without class. When the Europeans first came to Mexico, they discovered that the Aztecs had a very stratified society with noblemen at the top, and after the Spanish conquest, Cortez agreed to keep the Aztec social structure with the nobles above the peasants. As New Spain became more independent from Spain, the Spanish-born in New Spain did not just look at race to decide a person’s status in society. As Bakewell explains, “Broadly speaking, Spaniards in the Indies in the sixteenth century arranged themselves socially less and less by Iberian criteria or rank, and increasingly by new American standards. [...] simple wealth gained from using America’s human and natural resources soon became a strong influence on social standing” (160–163). After living in Mexico for over 25 years, I am of the opinion that a person could have darker skin and be considered high class either because they were from Aztec nobility or if they are very wealthy. This stratified society still exists today, with the majority of the country’s business class having lighter skin, in part because of the “castas” system. However, in politics, the indigenous population has been represented by Benito Juarez, who was elected to the highest level in 1861. He was the first Indigenous President of Mexico, born of the Zapoteco peoples in Oaxaca State. Mexico’s has a complicated history is reflected in its mixing of class and race and the founding national myth that Mexico is a ‘Mestizo’ society, a mixture of indigenous and European blood lines.

When discussing capitalism and how it gained control of Mexico’s political economy, Adam David Morton, describes it as a “[...] passive revolution [...], which is a [...] ruptural moment in the history of modern state formation in which capitalist development is instituted and/or expanded” (2011: 239), and this results in a post-colonial state whose development creates an uneven society and favours the ruling class. Morton also observes that the ruling families aided in the rise of capitalism in Mexico through keeping the racial and socioeconomic-based class levels, reinforcing the castas system and taking advantage of the workers. In *Revolution and State in Modern Mexico: The Political Economy of Uneven Development*, Morton refers to the Gramcian concept called ‘passive revolution’. He describes a passive revolution as “[...] a technique of statecraft that an emergent bourgeois class may deploy by drawing in subaltern social classes while establishing a new state on the basis of the constitution of capitalism. It is in this second sense that the outcome of the Mexican Revolution can be referred to as a period of passive revolution” (Morton 2011: 39). The Mexican revolution established the bourgeoisie as the dominant class; capitalist accumulation increased, giving them more resources and control (Morton 2011: 46–47). This created an environment where the bourgeoisie became very powerful and used its power to influence the country's population, as well as cultural and social values. This hegemony was achieved after the revolution, when the capitalist class established institutions that had a cross-class or corporatist character, such as state-sponsored unions and peasant organizations based on a clear class hierarchy under the control of the mostly light-skinned ruling class.

Cultural hegemony was developed in Mexico through specific families belonging to the ruling classes that held power though privately-owned corporations, and "[...] the conditions of hegemony were limited to privileged groups based on a central core of dominant classes" (Morton 2011: 125). The three families I am focusing on in this paper benefited from the Mexican revolution and the capitalist class it created, and took advantage of their position to control the 'subaltern social classes' through mass media communication, and more specifically through their usage of advertising messages.

The Families

In 1904, Juan Servitje Torrallardona immigrated to Mexico from the Catalan area of Spain. He worked in a Mexico City bakery called *La Flor De Mexico*. In 1926, he opened a business called El Molino. In 1936, Lorenzo, his oldest son, took over control of the business after he died. In 1945, Lorenzo and his family decided to industrialise the bread industry in Mexico and create a factory under the brand name of *Pan Bimbo*. Lorenzo said in an interview in 2006 that the secret to their success was "[...] to go to bed early, then wake up early, never stop working and to advertise [...]" (Patterson 2007: 254). In Forbes Magazine's annual survey of the richest billionaires in the world, Daniel Servitje is listed as number 310, and the corporation Bimbo is listed as the world's largest bread maker (Forbes 2012). In 1996, Lorenzo senior and his brother Roberto were behind the creation of the organization *A Favor De lo Mejor*, whose objective was to "[...] restrict the hours of exhibition of scenes of sex and violence on TV [...]" (Patterson 2007: 244). This was in reaction to the daytime talk show *Cristina* (on the Televisa network) that discussed current social issues such as teenage pregnancy and sex. They also tried to stop Madonna from performing in Mexico in 1993 because she supposedly "[...] masturbated on stage [...]" (Patterson 2007: 243–244). This organization was designed to remind Mexicans about family values. This is an example of cultural hegemony, with a ruling family trying to impose its morals onto a society by stopping public discussions on abortion, sexual education and condom use in Mexican mass media. Since the formation of modern Mexico, the blending of state and religion is not allowed. However, the Servitje family used its influence and bullied media outlets by threatening to pull their advertising from supposedly 'immoral' TV shows such as *Big Brother* and attempted to impose Catholic business views by controlling the way the media and advertising messages were constructed.

Eugenio Garza Lagüera is from a prominent Spanish family directly descended from Marcos Alonso de la Garza y Del Arcon, who ruled a part of Mexico on behalf of the Spanish royal family after 1596. Eugenio Garza Lagüera's grandfather,

Isaac Garza Garza, was one of the founders of the Cuauhtémoc Brewery in 1890, which later became Cervezceria Cuauhtémoc Montezuma. Eugenio's father, Eugenio Garza Sada, was one of the founders of the private University *TEC de Monterrey* and President of FEMSA (Fomento Económico Mexicano, S.A.B. de C. V), which is currently the largest bottler of Coca Cola in the world. Eugenio Garza Lagüera ran the company until his death in 2008, and now his widow Eva Gonda Rivera runs FEMSA along with members of the family (Forbes Femsa). In 2010, FEMSA sold its beer making operations to Heineken in exchange for a "[…] 12.5% stake in Heineken and 14.9% stake in its parent firm Heineken Holding" (BBC). The Garza Lagüera family still has a major say in beer operations in Mexico and influences Heineken globally because one of the daughters, Eva Maria Garza Laguera Gonda, is married to the current CEO of FEMSA, Jose Antonio Fernandez Carbajal (Forbes FEMSA). One of the beers in this company's holdings is Tecate, which has broadcast very controversial advertising in Mexico. One recent campaign offended the CNDH (The National Commission of Human Rights) with a billboard that portrayed three women at a bar with the headline "Buffet. It's easy to be a man. Tecate Beer, for you". The outdoor advertisements were taken down by the bottler only after a public outcry (CIMA) (TECATE) (Fig. 13.4). The current television ad "Choro" has also caused controversy; in 2014, it won a Publiviboras award, which was granted by an organization that recognizes the worst advertising campaigns promoting sexual violence, misogyny and stereotypes. (CNN) (Choro). I will perform a semiotic analysis of this commercial later on in this paper.

Alberto Bailleres Gonzalez's family is Mexico's third richest family, worth over $8.5 billion, as it owns one of the world's largest silver mines and the luxury department store chain Palacio de Hierro (Forbes 2015). In 1946, Raul decided to open a private university in Mexico called the *Instituto Technologica de Autonomous de Mexico* (I.T.A.M) because, in the words of its current Dean, one of the main Mexican public universities, UNAM, "[…] radicalized its graduates with

Fig. 13.4 Billboard exhibited in Mexico. Title "Buffet. It is easy being a man." Product: Tecate beer

Marxist-Leninist theories [...]" (Patterson: 131). This worried Raul Bailleres, who considered that "[...] the Mexican business class needed to be formed technically, taught by economists that excelled in finances" (Patterson 2007: 131). Alberto has been described as a man with many passions, including architecture, design and advertising. He was personally involved in the store's advertising campaigns and selected the "[...] aspirational concept for Palacio de Hierro: am totally Palacio [...]"; the campaign has always consisted of "[...] beautiful blonde white women [...]" (Patterson 125) since its inception to the present day.

As mentioned earlier, cultural hegemony in the Gramscian sense occurs when the ruling class influences the value system of a society so that the ruling class view is accepted as the norm by the rest of the society (Gramsci "Selections from the Prison Notebooks" 1971, "Prison Notebooks" 1992). This hegemony is currently being performed in Mexico through the ruling classes' unconscious persuasion of communications. The bourgeoisie maintain the cultural hegemony the Aztecs started and the Europeans built upon over 500 years ago, and today's society accepts the continued existence of a class system. Mexico's social and cultural values are influenced by the ruling families' powerful connections and ties: The Servitje family created an organization to impose family values and bullied media companies to accept the family's communications agenda; the Garza family created its own university, TEC de Monterrey, and paid for and approved misogynistic beer advertising that reduces the female to an object; and the Bailleres family has created its own anti-Marxist private university, ITAM, and imposed their personal social values by approving a long-running advertising campaign that promotes the white ideal lifestyle and reinforces the class structure.

Mass Control

Contemporary theorists such as semiologist Roland Barthes argue that social groups tend to regard as 'natural' whatever bestows privilege and power upon themselves. In *Mythologies*, Barthes analyzes semiotically how the bourgeois manipulate popular culture in their favour; he saw myth as serving the ideological interests of the bourgeoisie. In Barthes' book *S/Z*, he declares, "Bourgeois ideology turns culture into nature" (Barthes 1974: 206). He proposes that myths are an ideological way of creating acceptance through naturalization. They are designed to make ideological, social and cultural values seem perfectly natural. The advertising messages broadcast by brands can also be considered tools that allow the bourgeoisie to manipulate their myths into masquerading as truth by creating messages that contain codes or 'signs' that are broadcast to consumers. Through repetition, these myths can become accepted as truths, as in the case of religious conversion, which occurred in various parts of the globe, including Mexico, with the integration of new religious beliefs into the original mythic structures. After 500 years of colonial ideology and religion, many Mexicans are constantly faced with images and text placing lighter-skinned humans in positions of power. For example, lighter-skinned European saints replaced darker-skinned Aztec gods. The Catholic Church eliminated all evidence of the Aztec goddess Tonantzin (Aztec mother earth) and replaced her with the Virgin de Guadalupe. The Aztecs had built a temple dedicated to Tonantzin on a hill in Mexico City. After arriving in Mexico City, the Spanish destroyed the temple, and according to legend, the Virgin de Guadalupe appeared. To this day, the Mexicans still refer to the Virgin as Tonantzin. I propose that this myth was designed to bring the Catholic religion closer to the Mexican people and convert them to Catholicism (Figs. 13.5 and 13.6).

In his book *Propaganda*, Edward Bernays states that in order to have a democracy, the masses need to be controlled. He explores the psychology behind manipulating masses and the ability to use symbolic action and propaganda to influence politics, effect social change, and lobby for gender and racial equality. By keeping the masses distracted by frivolous events and relatively unimportant needs, consumers will not interfere with the activities of what he called 'the important few' (1927: 4). He called this scientific technique of opinion-moulding "the engineering of consent" and wrote a book on it by the same name in 1955. I argue that through the lens of Bernay's writings, the ruling families' current appropriation of advertising messaging is a tool to maintain power, because mass media can be manipulated to broadcast social and cultural values to the masses to reinforce perceived concepts already instilled in the Aztec society and later built upon by the Mexican society created after the Spanish conquest.

The concept of 'the gaze' provides another means of manipulating the masses through mass media. It was first written about by feminist film theorist Laura Mulvey in the 1970s in her article *Visual Pleasure and Narrative Cinema*, which discusses the "male gaze" and how the female body was objectified through the

Fig. 13.5 Tonantzin 2006 by Carl Jones. Digital ink and paint on canvas

Fig. 13.6 Virgin de Guadalupe. Oil on material. Created around 1531

medium of film because of male dominance in the movie industry (1975: 6–18). In the late 18th century, Jeremy Bentham designed the panopticon, a circular prison with a control tower in the middle; this allowed for more effective visual surveillance of the inmates, giving the guards more power, control and knowledge over the prisoners because of the 'panoptic gaze' the architecture offered them. In his book *Discipline and Punishment* (1975), Foucault relates the panoptic gaze to power relationships through knowledge and visuality. In *Practices of Looking*, Sturken and Cartwright state, "The gaze is integral to systems of power and ideas about knowledge" (2009: 103). In today's technologically-dominant world, the gaze creates a power dynamic where the consumer—in the quest for knowledge—allows electronic devices, such as a television, to deliver signs—in the form of entertainment—that feed the capitalist system's need for commodities. I think of the television as being a panopticon that spreads information through a visual and audio medium to control the masses. Television has become one of the devices appropriated by the ruling classes globally to disseminate information to consumers, who then reciprocate by purchasing advertised branded products, thereby giving money to companies that are owned by the ruling class, who then create new brands to sell to the consumer. The signs used in the messaging not only reflect, but also construct, social and cultural values (Shaheed 2014: 14).

Signs

Semiotics is an important area of study, and its theories can be applied to communication. One of the fathers of semiotics is Swiss French, Fernand de Saussure, who in the late c1800 thought that communication systems, particularly natural languages were not there to simply classify things as part of a reality, but they had a social aspect, reflected in the way they were structured. He stated that signs were compromised of two elements: the signifier and the signified. The signifier carried the message and the signified was communicated, however, he considered that the sign was completely discretionary, and that there was not necessarily a connection between the sign and its meaning. With many media channels broadcasting over 5000 messages a day to the typical urban consumer (Walker-Smith 2006), the studies and observations of theorist Roland Barthes are very important when applied to messaging that is broadcast through mass media. Barthes' application of the Saussurian model of 'signifier + signified = sign' to advertising can bring insights such as 'myth creation' into symbolic messaging applied to global brands, such as Marlboro, that are instrumental in exporting capitalism to non-Western nation states through globalization (Barthes "*Mythologies*" 1972: 115). The cowboy myth for Marlboro cigarettes is created through positive signs that have been removed from their original context, such as the Amerindian genocide. A sign can consist of symbols, words, images, or sounds and "[…] *is used in semiotics to encompass anything that stands for something other than itself*" (Danesi "*Encyclopaedia of Media and Communication*" 2013: 590). Barthes notes that semiological signs "appear […] as soon as there is a society, every usage is converted into a sign of itself" and this is because of "semantization" (1972 "Elements of Semiology" p. 41). Where society assigns a meaning to an object, and that particular meaning can be changed or re-interpreted through constructed advertising messages. Semiotics allows academics and professionals to "[…] compare and contrast objects from two different semiotic systems—language and imagery—and make a valid, useful analysis" (Harrison 2003: 48). Advertising research companies such as Marketing Semiotics Inc., based in Chicago, apply semiotic analysis to global brands such as American Express, McDonald's and Kraft Foods, so that they can successfully position their products in the marketplace.

In his book *Mythologies* one of the themes Barthes explores is how popular culture expressed through mass media contains messages that are composed of signs that can be appropriated and re-interpreted to construct myths that benefit the bourgeoisie society. I argue that Barthes exploration has inspired the new area of study called Visual social semiotics, which is defined by Jewitt and Oyama as covering "[…] the description of semiotic resources, what can be said and done with images (and other visual means of communication) and how the things people say and do with images can be interpreted" (Jewitt 2001: 136). It is important to note that advertising messages are created by communicators, and they have power and control over the signs and how they are combined to create unique messages that can change the original intent of the signs used. As Harrison states in her article "Visual Social Semiotics: Understanding How Still Images Make Meaning", this manipulation of signs "can [...] therefore affect and even alter meanings" (Harrison 2003: 40). In semiotics, it is believed that people view the world through signs, and as Chandler points out, "[...] although things may exist independently of signs, we know them only through the mediation of signs. We see only what our sign systems allow us to see". He explains further: "Semioticians argue that signs are related to the signified by social conventions in our use of various media that they seem 'natural', and it can be difficult for us to realise the conventional nature of such relationships" (2001: http://visual-memory.co.uk/daniel/Documents/S4B/). I am using visual social semiotics as a tool to analyze television commercials, which are messages composed through combining visual and audio signs to construct a 30–60-s narrative.

Media Semiotics

The sponsored messages that are produced and broadcast by the ruling families are texts composed of signs and codes, blended through the combination of words and images, and expressed in the most accessible mass medium—television—in order to affect Mexicans. Danesi states that "[...] brands are one of the most important modes of communication in the modern media environment" (Danesi "Brands" 2006: 3). In the book Understanding Media Semiotics, Danesi observes that television is a "social text" that creates a *cognitive compression effect* "where [...] viewers have little time to reflect on the topics, implications, and meanings contained in its messages ...TV images themselves do the thinking for the viewer" (Danesi 2002b: 145). Visuals and sounds together create a message in a television commercial, and because of the time restraints of 30 or 60 s, the images and audio are edited with precision to deliver a branded message in a short amount of time. Over time, "[...] people are affected by media images mainly because they reflect or reinforce already established trends within the culture" (2002: 151), and since the Mexican population has been receiving messages reflecting class structure and preferences for lighter skin, the advertising messages reinforce these social and cultural values. A television commercial is constructed through a collection of consecutive images that create a narrative. A television commercial controls the eye of the consumer, and it can be considered a way to control the gaze of the viewer in order to construct a specific narrative. This technique is similar to a mural painting, where the eye is directed towards various 'scenes' that together construct a narrative. This is also the case with commercials: The individual scenes tell a small story through the combination of signs and codes, and collectively, they tell a complete narrative that the consumer is expected to piece together. This method of presenting a narrative in the form of a mural is a visual technique appropriated by the Mexican government who hired muralists such as Diego Rivera or Orozco to depict important moments in Mexican history that were broadcast to the public. Television controls the viewer's gaze and construction of the narrative, instead of allowing the viewer to manufacture the scenes in any order. A television commercial can be considered a gaze that is stitched together, or a sutured gaze, because the audience places itself into the message (Miller 1977) so that it can understand the narrative.

Advertising

The ground-breaking 2014 United Nations (UN) report on the impact of advertising and marketing on cultural practices reveals some important observations on how paid communications can influence consumers' perceptions of themselves through the use of coded messages, composed of signs and symbols.

> Advertising and marketing strategies have become more sophisticated and brands have developed their own identities. Using a combination of meanings, symbols and values and having unmatched outreach worldwide, they provide codified messages to people and have succeeded in becoming some of the reference points for people's perceptions about themselves, others and the world in general. (Shaheed 2014: 14)

The UN committee observes that, in economic systems like the free market, consumers of brands do not make their own choices, but "[...] *instead, they operate within a sociocultural, economic and political framework that shapes and limits how they think, feel and act in the contemporary marketplace. Advertising and marketing practices increasingly help to shape this framework*" (Shaheed 2014: 15). This UN report's findings demonstrate that advertising does control how the consumer moves within their sociocultural world and that marketers are aware of how brands can influence the consumers' thought processes.

Advertising messages are composed of words and images and are created for specific media broadcasting a visual message sometimes complemented by audio or image only. The manufactured communication that consumers read is a "visual language" that information-mapping pioneer Robert Horn describes as "[...] *the tight coupling of words, images, and shapes into a unified communication unit. 'Tight coupling' means that you cannot remove the words or the images or shapes from a piece of visual language without destroying or radically diminishing the meaning a reader can obtain from it*" (Horn 1999: 27). Advertising is created to deliver a specific brand message; however, there are other messages sent out by an advertisement to the consumer that the consumer does not consciously process.

13.2 The Study

In the following study, I am exploring how Mexican sign conventions are appropriated and re-arranged to create branded messages that benefit the three ruling families in Mexico, unbeknownst to the audience.

Setting

The study was conducted within Mexico's capitalist economy. It analyzes branded messages that were first broadcast in 2014 on Mexican privately-owned networks such as Televisa and TV Azteca. The three brands selected each represent a company owned or part-owned by a ruling family based in Mexico. The advertisements were obtained from the *Circulo Creativo* (Creative Circle), the only organization that represents the creative knowledge workers in Mexico's advertising industry. I was president of this self-governed association during 2000–2001. Every year, a competition called *El Circulo de* Oro (Golden Circle) is held, where advertising and design agencies enter branded messages from various media to compete for gold, silver and bronze medals (www.circulocreativo.com.mx). This competition is held in order to select the most creative advertising messages broadcast within the borders of Mexico during a specific time period and to receive industry recognition from peers.

Data Collection

I approached the Creative Circle during May 2015 to obtain access to over 1400 advertising messages entered in the most recent *Circulo de Oro*, held in Acapulco in December of 2014. After reviewing all the entries, I selected three television commercials, each representing a branded commodity and owned by each of the three ruling families that I have chosen to examine in this paper. The current president of the *Circulo Creativo* Ivan Carrasco, Vice President Creative Director of Ogilvy and Mather Advertising, based in Mexico City, handed over this data.

Limited Data set

Modified grounded theory, visual social semiotics and statistical strategies were involved to analyze the data. I started the investigation using methods discussed in *Qualitative Communication Research Methods* by Lindlof and Taylor.

(1) I used the method of "open coding" through the "unrestricted coding of data" after categories were "built and then named" (Lindlof 2011: 250–251).
(2) I created a "codebook", which is "a tool for the development and evolution of a coding system and is an important means for documenting the codes and the procedures for applying them" (Lindlof 2011: 250–251). I constructed a codebook in the form of observational charts created for each commercial, while searching for emergent themes, so that I could develop, apply, revise and confirm a relevant coding structure. Each code is defined and illustrated with examples.

The three codebooks consist of one television commercial aired during the last year for one Mexican brand owned by each family. A visual social semiotic analysis of each commercial is performed by first creating categories based on skin colour, social class, gestures, text tracks and gaze.

SKIN COLOUR: For the coding of skin colour, the globally recognized *Fitzpatrick Phototyping Scale* was selected (Fitzpatrick 33–34). This scale recognizes differences in skin colour rather than specific racial characteristics, which tend to be socially-constructed concepts (Sachdeva 2009). I would like to note that the Fitzpatrick chart is not perfect in its design, as it places white skin at the top of the scale and ends with black (Fig. 13.7).

According to the CIA (Central Intelligence Agency) of the US government, 10% of the Mexican population is considered "mostly European", 62% Mestizo (mixed European and Amerindian), 21% "predominantly Amerindian", and 7% "Amerindian" (CIA 2015). For the purpose of this study, I am placing "Mestizo" and "predominantly Amerindian" together to represent the majority of the Mexican

Fig. 13.7 Fitzpatrick phototyping scale. Produced by Australian Government. Australian Radiation Protection and Nuclear Safety Agency. http://www.arpansa.gov.au/pubs/RadiationProtection/FitzpatrickSkinType.pdf. (Fitzpatrick 1975)

population, which totals 83%. This is a view supported by anthropologist Guillermo Bonfil Batalla who, in *Mexico Profundo*, proposed the idea that the remaining Amerindian communities—the 'de-indianized' rural Mestizos and large areas of the urban poor—make up a *Mexico Profundo*, where their lives and culture are based in the Mesoamerican civilization (Bonfil). It is important to note that the CIA report mentions that "Mexico does not collect census data on ethnicity [...]" (2015). Using those statistics, I am assuming that "European" is Fitzpatrick Type I and II and that "Mestizo, and predominantly Amerindian" is Fitzpatrick Type III and IV (Table 13.1).

According to the 2012 Hernandez-Zarate medical study of skin cancer patients, the majority of the Mexican population moves between Fitzpatrick Type III and IV, with 44% of population being between Fitzpatrick Type III and 34% being Type IV, for a total of 78%. In the dermatological study (Hernández-Zárate et al. 2012: 33–34) and in the CIA ethnic study, they would be classified as "Mestizo & Predominantly Amerindian", with a total of 83% of the Mexican population. The number in the dermatological study is 78%, and the CIA ethnic study is 83%—they are similar enough to justify the percentages of non-white, mixed-blood peoples in the Mexican population.

SOCIAL CLASS: Due to Mexico's history, I have created a code that deals with class and class structure. Class is another form of dividing and classifying Mexican citizens. This division had been performed by the Aztecs in Tenochtitlan, and then by the Spanish. Cultural codes such as social class are projected by the accumulation of signs that are constructed to present a collective message of "having class", which is achieved through the purchase of branded commodities like watches, clothing, shoes, furniture and cars. This collection of signs gives agency to the subject in the manufacture of their identity. In this codebook, I am observing codes and signs to analyze how individuals construct their identity to reflect their class.

In advertising, when a commercial is designed to appeal to a specific target group, the 'social economic classes' are reflected in levels such as A and A+, which are the top 1% of the population; B and B+ are middle to upper middle class; C and C+ are lower middle class; D is low class; and E represents the poorest consumers. The socioeconomic classes are categorized by the Mexican government according to the information collected and classified by the *National Institute of Statistics and*

Table 13.1 Chart presenting the results of the Hernández-Zárate et al. (2012), Fitzpatrick Scale and CIA racial breakdown of Mexican population (2015)

Racial skin types in dermatology study Hernández-Zárate	Fitzpatrick scale	CIA definitions	CIA Mexican population (%)
N/A	Type I and II	"Mostly European"	10
44 and 34%	Type III and IV	"Mestizo," "Predominantly Amerindian"	83
N/A	Type IV	"Amerindian"	7

Geography (INEGI) (INEGI 2009) and are based on 13 variables: (1) education level of head of household, (2) number of rooms, (3) number of bathrooms with shower (4) type of floor, (5) number of light bulbs, (6) car, (7) boiler or water heater, (8) automatic washing machine, (9) videocassette player, (10) toaster, (11) electric vacuum, (12) microwave, and (13) personal computer.

GESTURES and body language can be considered a cultural code and reflect thought, emotions and attitudes that are spoken and unspoken by actors. The codes for the gestures presented in the commercials are designed to examine how body language and the use of signs in the form of props accentuate the gestures and assist in presenting the narrative.

TEXT TRACKS such as words, music, and audio are heard or read in the branded television commercial. Text tracks can represent signs and cultural codes, such as

- social status, conflicts, problems and solutions, and advance the narrative. Word patterns can be observed in both written and oral formats, and audio patterns can be heard through music and sound cues.

GAZE: Signs can be implied through eye movement, eye direction and camera angles. The camera can substitute for the viewer's gaze and broadcast signs as to how the content should be seen. The camera also takes the place of a fourth wall, the wall that is never seen on the screen. The actual gaze of the actors in the commercial can give clues to cultural codes and give meaning to a narrative.

(3) A "syntagmatic analysis" was performed to expose the semiotic messaging broadcast to the Mexican public. "Syntagmatic analysis" involves analysis of a text by "studying its structure and the relationships between its parts" (Chandler) in order to discover the representational metafunction behind the manufactured image. According to Harrison in *Visual social semiotics: Understanding how images make meaning*, the representational metafunction is "[…] about the people, places, and objects within an image—the represented participants—and answers the question 'What is the picture about?" (Harrison 2003: 55). I will bring together my conclusions from the five codes of skin type, social class, gesture, text tracks, and gaze to analyze the pictures or scenes that compose the television commercial.

13.3 Results and Analysis

Advertisement 1

Company: Bimbo
Name of family that owns company: Servitje
Client: Barcel
Advertising Agency: Publicis Mexico

Product: "Runners" Chip snack
Title: Monsters
Length: 27 s
Medium: Television
Year created and exposed on paid media: 2014
Link to commercial: https://vimeo.com/136048999 (password is: Mexico)
https://www.youtube.com/watch?v=9PPJXzBuMw4
Actors: Four main actors in commercial.

Company

Barcel is a sub division of "Grupo Bimbo, the world's largest bread maker, which had 2009 revenues of US$8.6 billion" (Forbes 2012). Barcel makes salty snacks directed at the general market. Its main competition is Frito-Lay, owned by the American corporation Pepsi-Co. I have worked on advertising for both Bimbo and Pepsi-Co.

Fig. 13.8 (A) 1 Male, approx. 15 years of age (Fig. 13.8)

Fig. 13.9 (B) 1 Male, approx. 14 years of age (Fig. 13.9)

Fig. 13.10 (C) 1 female, approx. 15 years of age (Fig. 13.10)

Fig. 13.11 (D) 1 adult male, approx. 35 years of age (Fig. 13.11)

Narrative

Commercial starts with 3 teenage students in front of high school eating Runners brand snack chips. A large car from the late 20th century arrives and parks in front of them. The students tell him not to park there because they are saving it for someone. A 35-year-old man steps out of the car. Cut to a close-up of the 3 students, now wearing crash helmets. A 'Runners' monster truck drives into the frame, smashes into the car, crushing it. The man falls to the ground in shock. Cut to 10 people cheering, holding a sign that says "Adrenalin". Cut to wide shot of monster truck in a stadium; its wheels are spitting out "Runners" chips towards camera and 2 Runners-branded packages appear.

Fig. 13.12 (E) 10 supporting actors (5 females and 5 males), ranging from 18 to 65 years of age (Fig. 13.12)

Fig. 13.13 (F) Driver of Monster Truck (Fig. 13.13)

Skin classification in television commercial

Decoding the skin types

In this commercial,10 actors move between Fitzpatrick Type I and Type II, representing 66.67% of the actors in the commercial. However, in the actual demographics the "Europeans" or Skin Type I and II make up 10% of the Mexican population. Only 5 actors representing 33.33% of the people in the commercial are in the Fitzpatrick scale of III and IV, representing the typical ranges of "Mestizo, predominantly Amerindian" skin colour. However, in the real-life demographics, Types III and IV make up 80% of the Mexican population. There are no

Table 13.2 Code: skin type

Skin type descriptive Fitzpatrick scale	Definition	Examples
Type I	Always burns, never tans(pale white skin)	Actors E (2 background actors)
Type II	Always burns easily, tans minimally (white skin)	Actors A, C, D and E (4 background actors)
Type III	Burns moderately, tans uniformly (light brown skin)	Actors B, D, and E (3 background actors)
Type IV	Burns minimally, always tans well (moderate brown skin)	E (1 background actor)
Type V	Rarely burns, tans profusely (dark brown skin)	
Type VI	Never burns (deeply pigmented dark brown to black skin)	

"Amerindian" skin types in the commercial, so 7% of the population are not portrayed. The commercial does not represent the majority of the skin Types III and IV of the Mexican population (Table 13.2).

Class description in advertising

Decoding the class signs

The group of student actors—A, B, C—are wearing clothing that is currently popular with their age group in Mexico City suburbs, and it appears that they are from class levels C+, B, and B+. They are wearing the latest student fashions and student C (girl) has expensive accessories, including a plaid bag and large Beats by Dr Dre branded headphones that attach to a cell phone. These class-coded accessories would place her in the upper middle class. The crowd, E, wears clothing that represent current fashion trends indicating they are from mid-class levels. The car driver, D, is wearing items that represent the signs of being from a lower class level. He has accessories that consist of a gold necklace and bracelet—items that people from the upper levels would consider flashy and tacky. Under his top, he is wearing a "wife beater" shirt. The mullet hairstyle is a throwback to the 1970s and is out of fashion. His clothing is not coordinated, and the fact he is wearing long socks with shoes and shorts further classify the car driver as a "naco", which is a term used in Mexico to classify people who are from less-educated the lower classes, and they would be classified as level D. Age differences between the car driver and teenagers are also emphasised through the clothing styles between the car driver—who wears clothing that was popular last century—and the students, who sport current fashion trends and technology. The helmets that the students put on after confronting the driver, is the same brand of helmet that the monster truck driver wears as he crushes the "naco" driver's car. The monster truck is an expensive contemporary vehicle whose driver wears a futuristic black helmet, which contrasts against the cheaper, old-fashioned car and driver, further emphasizing differences between the two classes. The black helmet can be perceived as protection from an accident, or a way of blocking sound or sight from entering (Table 13.3).

My analysis is that the helmet represents protection for the bourgeoisie middle classes against the lower class "nacos", and the only way to obtain that protection is by buying the snack Runners that is distributed by the monster truck. The commercial contains signs and codes that reinforce socioeconomic class differences and demonstrates the need for the bourgeoisie to be protected from the lower classes, implying that they can protect themselves through consumption.

Table 13.3 Code: social class

Class descriptive	Definition	Example
Clothing	Attire worn by actor	Actor A Red t-shirt with straight leg faded blue jeans Actor B Stripped t-shirt covered by short-sleeved shirt with blue army camouflage design, straight leg faded grey pants Actor C Long-sleeved sweatshirt; cut-off jean shorts Actor D 'Wife beater' shirt covered by a white short sleeved Hawaiian shirt; Long wide light green sport shorts; long white knee sports socks; regular shoes with bright laces Actors E 10 background actors wearing spring-like clothing featuring blue, light green and light browns colours Actor F Black gloves and black leather top
Accessories	Props held or possessed by actor(s) in ad	Actors A, B and C have Runners packages in their hands. At the 10th second of the commercial black cycle helmets (similar to the helmets Daft Punk wear) appear on heads Actor C Plaid shoulder bag; Beats by Dr. Dre type headphones Actor D Gold chain around neck; gold bracelet on left wrist; 1980s pale yellow cardoor has been modified to open as luxury car Delorean; rabbit foot dangles off rearview mirror Actor D Black Helmet
Hairstyle	Hairstyle and facial hair	Actor A and B have hair cut around ears in contemporary style Actor C has long shoulder length hair Actor D has an exaggerated mullet hairstyle with excessive length at back Actor D has a moustache and is unshaven

Gestures/body language descriptive of each actors 'performance' in advertising

Decoding the gesture signs

The actor D uses aggressive physical gestures negatively throughout the commercial to communicate with the students. He raises his arms in a challenge when he is told not to park in the 'saved' space. He also uses his arms and turns his body to 'magically' close the door and turn off the car's music. The students express their feeling in a subtle, passive manner, shaking heads, showing disgust and putting on black helmets to protect themselves from what is to come (Table 13.4).

The crowd E is very expressive in their body language, demonstrating extreme happiness when the car of actor D is smashed by the monster truck. The students and crowd E together are acting in a passive manner, while the actor D is

Table 13.4 Code: gestures

Gesture descriptive	Definition	Example
Negative	Not welcome here	Actor C Gestures with a shake of hand when actor D arrives Actor D Puts head out side of car and gestures "who cares" to actors A, B, and C
Confidence	Look at me	Actor D Does an exaggerated body move to bring attention to the door closing Body action appears to shut off car music
Confrontation	Between students and car driver	Actor C mentions to Driver D that it is taken "No, No, No. Esta apartado por un amigo" Actor D Parks car in a space that is already spoken for; he laughs at their comment Actors A B, C Look at each other in confusion and disagreement Actor D Rubs hand along car in a suggestive manner Actor F Driver of truck clenches hands around steering wheel of monster truck
Aggression	Driving monster truck	Actor D lifts head in a 'who cares' attitude. Arms risen on either side of his body in an aggressive stance Actor F Drives monster truck onto parked car
Upset	Not happy with what is/has happened	Actors A, B, and C when car takes parking spot Actor C Female student gives look of disgust to camera when actor D makes a suggestive gesture by running his hand along roof of car Actor D Drops to his knees when car is destroyed
Happy	Crowd watching the activities on the street	Actors E Crowd cheers when D's car is destroyed and he is on his knees on street
Preparation	Wearing of helmets (prop)	Actors A, B, and C wear helmets as a response to actor D's aggression Actor F Wears a helmet Actor Curls fingers around steering wheel

aggressive. I conclude that the brand demonstrates its dislike for overt aggressive gestures that lower classes are aggressive, and shows its support for a passive attitude that is attractive to the majority of Mexicans.

Text tracks in the commercial

Script:
SFX—Cumbria music in background grows louder as car approaches.
Actor C (Girl)—*No, no, no. Esta apartado por un amigo. (No, no, no. It is reserved for a friend.)*
Actor B (Boy)—*No (low voice).*
SFX—D Door shuts and music turns off.
Actor D (Man)—*Laughs.*
Actor F turns on Monster truck engine.
ANNCR *Llego todo la adrenalina a de los nuevos Monster Runners. (It has arrived, all the adrenalin of new Monster Runners)*
SFX—(Metal Crashing) Truck landing on car and crushing it.
ANNCR—*Nos costa la adrenalina. (All it takes is adrenalin.)*
ANNCR—*Sabour a turbo queso. (Tastes of turbo cheese.)*
Crowd—Cheers in reaction to previous scene.
SFX—Heavy metal guitar coincides with cheese Runners shooting out of the Monster truck's tires (Table 13.5).

Table 13.5 Code: text tracks

Text tracks descriptive	Definition	Example
Negative	Words/sounds/music used in a sentence that has a negative connotation Aggressive in tone	Actor C (Girl)—No, no, no. It is reserved for a friend Actor B (Boy)—No (low voice) Actor D (Man)—Laughs SFX—(Metal Crashing)
Positive	Words/sounds/music used in sentences that are favourable Happy in tone	Actors E Crowd—Cheers! ANNCR—All it takes is adrenalin SFX—Heavy metal guitar music that coincides with cheese Runners shot out of the Monster truck's tires
Problem	Words/sounds/music that indicate a problem	Actor C (Girl)—No, no, no. It is reserved for a friend SFX—Cumbia music in background grows louder as car approaches
Solution	Words/sounds used are favourable and part of the solution to the problem	Actor F revs Monster truck ANNCR—(It has arrived, all the adrenalin of new Monster Runners) ANNCR—Tastes of turbo cheese SFX—Heavy Metal music when car is crushed

Decoding the text

The word "No" is said 4 times by students A, B, and C to the car driver D. He responds by laughing sinisterly. The crowd E cheers after the car is crushed and a heavy metal riff is heard, representing the adrenalin of the monster truck coming and hitting the car. The upper classes usually listen to foreign music because they understand English lyrics. Imported contemporary popular music such as heavy metal is accessed through online services like iTunes, offered on expensive electronic devices such as iPhones—which is a luxury the less fortunate cannot afford.

The Cumbia music is heard when the car driver parks the car in a space that is not his. Cumbia is a very traditional music, and is usually listened to by the lower classes, as it is made available in accessible formats such as locally made CD's and even cassettes. Cumbia music has a traditional base in colonial Latin America, so Mexicans who want to get in touch with their latin roots gravitate to this type of music. However, it should be mentioned that all Mexicans of all socioeconomic level do appreciate cumbia music during traditional celebrations like weddings and special holidays.

There are a lot of binary cues used in the audio construction of the narrative. The "No" from the students; the "Ha, ha, ha" laughter from the driver; young age versus old age actor D; the cry of the man D when his car is crushed versus the cheering of the crowd; old car versus a monster truck; and, the cumbia music versus the heavy metal music.

There is a music war when the imported, contemporary heavy metal music overpowers the traditional cumbia music. This is demonstrated by the heavy metal music being played when the monster truck is introduced and when it crushes the car that emitted the cumbia music. Therefore, the heavy metal music is associated with victory, modernity and the upper classes, and the cumbia music is associated with losing, tradition and the lower classes.

Gaze

Decoding the gaze

The same camera angle is used throughout the commercial to present a suggested neutral narrative. However, the visual narrative breaks when a lower camera angle is used to demonstrate the joy of the crowd when the car of actor D is crushed. The gaze of the camera is shot from below and this helps to present the crowd as a hero, forcing the viewer to look up at the crowd. This technique of looking up to present a hero was used by Leni Riefenstahl when she presented images of the Nazi party in her movie *Triumph of the Will* (Triumph 1935) (Table 13.6).

Conclusion:

Decoding the signs and cultural codes

In his book *Mythologies*, Roland Barthes explains the concept of connotation and denotation—where myth occurs when the consumer reads connotative meanings as denotative. This concept, applied to the helmet visual device as seen in the

Table 13.6 Code: gaze

Gaze descriptive	Definition	Example
Eye level of viewer	Neutral framing of scenes. No opinion is given	Actors A, B and C in conversations with (D): Camera angles are from eye level
Camera angle from below looking up	Hero. Crowd watching the activities on the street	Crowd E Camera is placed at a lower level looking up to subjects

television commercial, produces the following connotative meaning: the snack Runners is actually a device to protect the young, mostly light-skinned consumer against the older lower class. What this signifies is that a food made from corn is actually a helmet that separates the upper class from the lower class and turns them into heroes. This meaning is socially and culturally specific and not a natural association.

This commercial demonstrates that during a confrontation, the higher class has power over the lower class. The helmet is a representation of the brand Runners and demonstrates how it will protect the young, upper middle class, mostly lighter-skinned students in their fight against the older, lower-class "nacos." The students also have the support of the mostly white majority. This is reinforced through the use of signs and cultural codes that emphasise differences in socioeconomic status, skin type, gestures, artefacts, age and music. Runners creates a mythical story of struggle between the contemporary bourgeois middle and upper classes versus the traditional lower class.

Advertisement 2

Company: Palacio de Hierro
Name of family that owns company: Bailleres.
Client: Palacio de Hierro
Advertising Agency: Teran/TBWA Mexico City.
Product: Department store image advertising
Title: "Andres Velencoso"*
Length: 60 s
Medium: Television and Cinema.
Year created and exposed on paid media: aired early 2015.
Link to commercial: https://vimeo.com/136048998 (password is Mexico)
https://www.youtube.com/watch?v=-HMzETI44BI
**TV Commercial not sourced from Circulo de Oro competition 2014.*
Actors: Two main actors in commercial.

Company

Palacio de Hierro is a department store that sells luxury brands to the A and A+ markets in major urban centers in Mexico. This is the description on its website: "El Palacio de Hierro is the most prestigious department store in Mexico, offering customers products and services of the most prestigious brands". The owner,

Fig. 13.14 (A) Actor, male 35 years of age. Name: Andres Velencoso. Model from Spain. Born March 11, 1978. Has appeared in many advertising campaigns (e.g. Chanel and Louis Vuitton) (Fig. 13.14)

Fig. 13.15 (B) Actor, female. Approx. 25 years of age (Fig. 13.15)

Alberto Baillares, has been described as a man with many passions, including architecture, design and advertising.

Narrative

The television commercial starts with scenes introducing the well-known Spanish model, Andres Velencoso, in full body shots and close-ups in various urban locations, during different times of the day and night. It is filmed as if it were a print fashion shoot, with scenes that have limited movement. A young female is introduced into the narrative as a voyeur observing him undressing. She finds her way to Velencoso and they slow dance. He takes her for a ride in a sports car, and for the rest of the commercial, we see Andres displaying various positive emotions directly towards the camera. During the advertisement, typography appears over scenes—"soy…" (I am)—combined with various positive adjectives. The commercial ends with the text "Soy totalmente Palacio" (I am totally Palacio]) and a large logo saying "Palacio de Hierro" (Palace of Iron]).

Skin classification in television commercial

Decoding Skin

I would like to note that the actor Velencoso does appear to have a Fitzpatrick Type III in most of the photographs that he appears in (Fig. 13.16); however, for this specific TV commercial and accompanying billboard campaign, his skin is lighter and is Type II (Fig. 13.14). His facial and body skin colour are the same: Fitzpatrick Type II (Fig. 13.17). The female (B) has Fitzpatrick Type I skin. Her facial skin type is the same type as her arms and legs (Table 13.7).

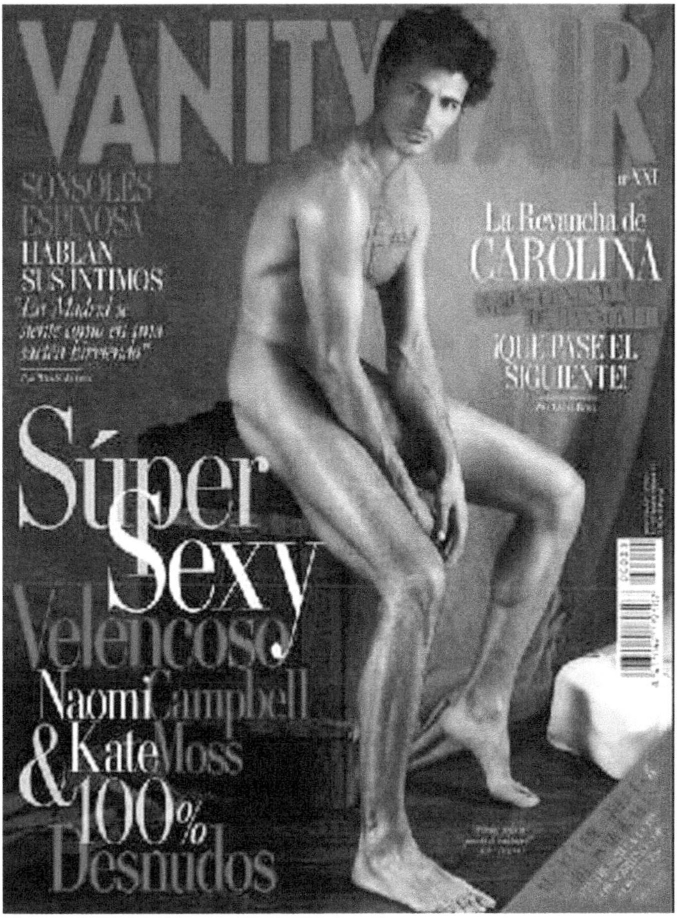

Fig. 13.16 Vanity Fair Magazine. Spain 2010

Fig. 13.17 Scene from TV Palacio De Hierro 2015

Table 13.7 Code: skin type

Skin type descriptive Fitzpatrick scale	Definition	Examples
Type I	Always burns, never tans (pale white skin)	Actor B
Type II	Always burns easily, tans minimally (white skin)	Actor A

Class description in advertising

Decoding Class

The male actor A wears what seems to be a new outfit in each scene, demonstrating various 'looks' and combinations of clothing that can be worn throughout the day and evening. The clothing, props, car and jewellery are all designer objects, representing the best-branded commodities that the department store has to offer. The latest trends from the fashion industry are designed to separate the higher classes from the lower through the use of coded signs. The commercial is also informing the consumer of the latest signs and codes that project the image of the upper ruling class (Table 13.8).

The female B appears as a luxury accessory to the male A to emphasise his seductive powers.

Gestures/body language descriptive of each actors 'performance' in advertising.

Decoding gestures

The main gestures observed seem to be that of actor A seducing B through his removal of clothes and overall use of his body. B demonstrates her attraction to A when she opens her mouth when she sees him removing his shirt through the open window. B makes herself available by walking around A when he is sitting in the chair. During the dance, A allows herself to be pulled tightly against his body. The male is in full control of the situation, using his masculinity to seduce the female and the audience (Table 13.9).

Text tracks in the commercial

The actors do not speak in the commercial; however, text appears during specific scenes and a soundtrack is present. It was music featured in the opening credits of the HBO series True Blood—"Baby did a bad thing" from Chris Isaac.

The script.

Text: QUIEN SOY? (WHO AM I)

Table 13.8 Code: class

Class descriptive	Definition	Example
Clothing	Description of attire worn by actor	Actor A All the designer clothing worn is sold in Palacio de Hierro. The actor wears suits, a tuxedo jacket, shirts, shoes, leather jackets, leather belts, various ties, bow tie Actor (B) The actor wears 2 different designer dresses, shoes
Accessories	Description of props held or possessed by actor(s)	Actor A wears a Breitling watch, drives Porsche sports car and sits on leather chairs. The female actor appears as an accessory in the car Actor (B) wears a few rings on her left hand

Table 13.9 Code: gestures

Gesture descriptive	Definition	Example
Seduction	Use of body language in free, open manner	Actor A Stands in front of glass window and removes shirt to reveal muscled chest Actor B Voyeur looking through window at male removing shirt Actor B Opens mouth when sees male removing clothing Actor A Sits confidently in chair while (B) walks around him, with her hand running along chair Actor B Allows herself to be grabbed from behind her back and pulled towards (A)
Confidence	Movement of body and extremities to attract attention	Actor A Looks directly at camera throughout most of the commercial Actor A Leans against walls in various situations, projecting confidence Actor A Sits in chair while female parades around him Actor A Pulls pelvis of (B) towards his lower body Actor B Sits on (A)'s lap while he stares at camera Actor A Drives car inside a stadium and does wheelies at a dock. He drives with confidence Actor A Dances by himself. (A) throws shirt at camera while he is shirtless Actor A Walks towards camera while taking off glasses with hand in pocket
Challenge	Body movement creates movement and action	Actor A Takes off shirt (A) Throws shirt at camera (A) Looks directly at camera, laughing (A) Runs towards camera through smoke

Text: SOY ATRACCION (I AM ATTRACTION).
Text: SOY ESTILO (I AM STYLE).
Music: Baby did a bad bad thing.
Text: SOY SEDUCCION (I AM SEDUCTION).
Music: Baby did a bad bad thing.
Baby did a bad bad thing, feel like crying, feel like crying.
Text: SOY EMOCION (I AM EMOTION).
Music: Oh, feel like crying, feel like crying.
Text: SOY LIBRE (I AM FREE).
Music: Oh, feel like crying, feel like crying.
Text: SOY AUTENTICO (I AM AUTHENTIC).
Text: SOY ASI (I AM LIKE THIS).
Text: SOY TOTALMENTE PALACIO (I AM TOTALLY PALACIO).

Table 13.10 Code: text track

Audio descriptive	Definition	Example
Statement	Adjectives used in text are favourable towards actor A	SOY ATRACCION (I AM ATTRACTION) Text: SOY ESTILO (I AM STYLE) SOY SEDUCCION (I AM SEDUCTION) SOY EMOCION. (I AM EMOTION) SOY LIBRE (I AM FREE) SOY AUTENTICO (I AM AUTHENTIC) SOY ASI (I AM LIKE THIS) SOY TOTALMENTE PALACIO (I AM TOTALLY PALACIO)
Music	Lyrics to accent visuals English-language music is imported and expensive to buy	Baby did a bad bad thing (repeated)
Typography	Selected fonts to represent an emotion or attitude	Sans serif, bold, aggressive font in upper case letters. It is an order telling the audience that "I AM AUTHENTIC". The same type of font used in the traffic sign STOP Logo Palacio de Hierro: cursive serif font

ANNCR: *Soy totalmente Palacio (I am totally palacio).*

Music: Guitar rift.

Text: Palacio de Hierro (logo), Visitanos: En tienda. En Linea. (Visit us in store, on line) (Table 13.10).

The lyrics are used to construct the narrative of a rich bad boy, suggesting that he is an outlaw. It is typical in Western culture to present a wealthy man who has done some "bad bad things". and, because of this, they are attractive to females. In the 1950s film *Rebel Without a Cause*, James Dean was an outsider whom women were attracted to, specifically because he projected an image of rebellion. In the more recent film *The Wolf on Wall Street*, the lead character attracted many women because of his bad boy reputation and wealth. In fact, American culture rewards rich men who behave badly. For instance, the owners of banks who caused the 2008 financial crash were not put in jail, and in 2009, many received bonuses from the banks they presided over. The music is in English and would be known by wealthy Mexicans who can afford to listen to English language music or watch pay cable, vs cumbia which is less expensive to access and cheaper to buy.

The commercial opens with a question—"QUIEN SOY? (WHO AM I)"—and then answers it with adjectives that are aspirational for most males. The typographic font is confident and bold. The commercial is structured so that the text is answered or reinforced with visuals. For example, when the word 'SOY SEDUCTION (I AM SEDUCTION)' appears, the actor A is "seducing" the female; when the words 'SOY LIBRE (I AM FREE)' appears, A is shown dancing by himself and doing wheelies with his car by the water's edge.

The commercial suggests that by purchasing imported branded commodities at Palacio de Hierro, any male can construct an ideal identity and become a member of the light-skinned ruling class.

Gaze

For most of the commercial, the same gaze is used to present a narrative that is from the point of view of the female, who is watching the male through the glass window. The audience thus becomes the voyeur as well, spying on the male. When the female has left the story, the audience continues to be a voyeur, and the actor returns the gaze. The actor A spends more than 50% of the commercial looking directly at the camera into the audience's gaze. For men, this can be considered confrontational. However, when the audience is put into the position of a "female" voyeur, it becomes a seductive gaze, trying to seduce the audience. The technique of inviting the audience into the commercial, of making them forget the camera and insert themselves into the moving image is called 'suture'. As explained by one of Lacan's disciples, Jaques-Alain Miller, "Suture [is], by extension—the general relation of lack to the structure of which it is an element, inasmuch as it implies the position of a taking-the-place-of" (Miller 1977 vol. 18, no. 4). Thus the audience is sutured into the commercial, allowing both sexes to be seduced by the powerful white male who represents the ruling class (Table 13.11).

Conclusion:

Decoding the signs and cultural codes

Applying Barthes' 1972 semiotic concept of connotative and denotative to the Palacio de Hierro television commercial produces the following denotative meaning: The male actor, who wears imported branded luxury products sold at Palacio de Hierro, becomes a hero. What this signifies connotative is that commodities such as plastic, cloth, leather, metal, and rubber—when worn together on a male body—can change him into an aspirational sex object. It is creating a branded myth. This meaning is socially and culturally specific and not a natural association.

Table 13.11 Code: gaze

Gaze descriptive	Definition	Example
Eye level of viewer	Neutral framing of scenes	The camera is at eye level It appears to be from the point of view of the female voyeur
Camera angle from below looking up	Hero	The camera does not look down upon the male It does look at eye level or up to the male
Gaze	Challenge	Actor B shown looking at actor A through a window as if she is a voyeur Actor A Spends most of the commercial placing his gaze directly at the viewer

The commercial is designed to manipulate men from any class level, skin type, or sexuality to willingly see through the eyes of a white female voyeur and be seduced by the actor and the brand. The commercial is a celebration of the A+ class's and Fitzpatrick skin Type I and skin Type II consumers' lifestyle. It demonstrates how this ruling class lives. It is a message that legitimizes the 'bad bad boys' of the ruling class and permits them to do what they want because they can. The commercial communicates that this lifestyle is obtainable to consumers—for a price. The ad creates a myth where the powerful white male becomes a hero, even though he is a "bad boy", and this is obtainable through the purchase of luxury goods. This constructed hero is able to seduce everyone he meets. The ad is manipulating the consumer into accepting and buying into the ruling class myth through the technique of a sutured gaze. As Foucault discusses in his book *Discipline and Punishment* about power relationships through knowledge and visuality, the powerful ruling families are using television to unconsciously spread information about the signs and codes of the ruling class to inspire the masses to spend money. This is accomplished through the audience's purchase of luxury commodities of businesses, which are owned by the ruling classes, who in turn use the mass media to construct social and cultural values. It is important to note that, for cultural hegemony to work, it does not have to be a conscious decision made by the ruling families; the signs and cultural codes in the advertising portray values which indirectly reinforce their class power.

Results and Analysis

Advertisement 3
Company: FEMSA
Name of family that owns company: Familia Eugenio Garza.
Client: Heineken, Cuauhtémoc Montezuma.
Advertising Agency: Nómades
Product: Tecate Beer
Title: Tecate Choro
Length: 1:27 s
Medium: Television
Year created and exposed on paid media: 2014.
Link to commercial: https://vimeo.com/136048997 (Password is: Mexico)
http://www.latinspots.com/site/sp/pieza/detalle/18440/comercial/Choro
Actors: Two main actors in commercial.

Company:

Eva Gonda, widow of Eugenio Garza and their 4 daughters (Forbes) own FEMSA and Cervezceria Cuauhtémoc Moctazuma. They sold the Cervezceria on January 11, 2010 to Heineken so that FEMSA ended up with a 20% stake in its parent firm, Heineken Holding. The objective was that FEMSA could have a larger international presence (BBC FEMSA) (femsa.com) and the sale gave them supervision of the Mexican and Latin American market and a position on the Heineken board.

Fig. 13.18 (A) Female approx. 27 years of age (Fig. 13.18)

Fig. 13.19 (B) Male approx. 35 years of age (Fig. 13.19)

The operations and marketing in Latin America is still under control of the Eugenio Garza's family, as Jose Antonio Fernandez Carbajal is president of LATAM Heineken committee (femsa.com).

Tecate beer was first created in 1944 and now is the flagship brand for Cuauhtémoc Montezuma and Heineken in Mexico.

The commercial is very long for a typical TV ad, which currently are 20 or 30 s long versus 20 years ago where they were 30 to 60 s. It was to be the first commercial shown before a World Cup match featuring Mexico. It was designed to have the names of the two teams playing as part of the information portrayed at the end of the commercial, and in this case it was Cameroon versus Mexico.

Fig. 13.20 (C) Extras. In the background (Fig. 13.20)

Fig. 13.21 (C) Extras. In the background (Fig. 13.21)

Narrative

The Television commercial opens with a romantic street scene—we see a couple walking upstairs to the door of a townhome. The female invites the male into the apartment, but he hesitates and explains that he wants to show her respect by not going in today. After a long explanation, the boyfriend leaves, and the girlfriend enters the home as he runs away. She is not aware that he went to watch the soccer match with his friends and was making up a reason not to go into her place.

Skin classification in television commercial

The male and female protagonists both have Fitzpatrick Type II skin. Of the 6 extras that appear, 5 are white-skinned and one is Type III. This demonstrates that the commercial features light-skinned actors and is not proportionately

Fig. 13.22 (C) Extras. In the background (Fig. 13.22)

representative of the skin type breakdown in Mexican society, which is 10% light skinned vs. 80% mid rage skin tones (Table 13.1).

Class description in advertising

The female actor A wears clothing that reflects an upper middle class and what she would wear for a daytime date. Her white top is Spanish in style and is reminiscent of the classic paintings where Spanish women are shown with a white top, revealing the shoulders. Actor B wears clothing that is appropriate for an afternoon date and that can be taken off to fit in with the guys watching a soccer game. The extras are wearing casual American-style clothing. It is not typical for Mexican males to wear shorts unless they are at the beach. Since the set is New York–style and the men are wearing casual American-style clothing, we can deduce that the location is not Mexico. This type of neighbourhood for a Mexican represents upper middle class because of the size of the buildings and the price of real estate—only a wealthy person could afford to live here. So Mexicans living in the world presented here would need to have a lot of money (Table 13.12).

Fig. 13.23 (C) Extras. In the background (Fig. 13.23)

After male B succeeds in convincing female A to let him leave, he runs from the situation towards a modern building. It is as if he is escaping romance and entering the current time period. The woman's house number is 69, which is the same number of a sexual position. This is not a coincidence. It establishes the female as a sexual object (Table 13.13).

Gestures/body language descriptive of each actors performance' in advertising

When commercial starts, B is standing at eye level with A. When B starts to explain why he will not come into the apartment, he stands a few steps below A. By standing below the female and looking up, the male gives the female a false sense of power in order to convince her that he is telling the truth. By standing on steps that are below where the female is, he is putting the female in a high position and forces her to look down on him, therefore appearing vulnerable. Instead of turning his head up, the male moves his eyes to look up, as a child would when speaking to their mother. The red rose is used as a device to represent the feelings A has towards B and vice versa. At the end of the narrative, A brings red rose close to her chest to represent love and acceptance (Table 13.14).

Table 13.12 Code: skin type

Skin type descriptive on Fitzpatrick scale	Definition	Examples
Type I	Always burns, never tans: pale white skin	Extras C × 1
Type II	Always burns easily, tans minimally (White skin)	Actor A. Actor B, Extras C × 4 of them with this level of skin
Type III	Burns moderately, tans uniformly (light brown skin)	Extra C 1 has this skin type

Table 13.13 Code: class

Class descriptive	Definition	Example
Clothing	Description of attire actors wear	Actor (A) Jean skirt, white embroidered top, belt, beige shoes with heels, and brown handbag Actor (B) Untucked grey blue jacket, light blue patterned shirt, jeans, brown shoes Extras (C) 2 × men in shorts. 1 × woman in pink top. Blue is the predominant colour. Casual street wear
Accessories	Description of props held or possessed by actor(s) in ad	Actor (A) Leather handbag, red rose One extra has a bike
Location	Description of set/house exterior	Actor (A) Lives in an upscale neighbourhood. There are flowers in pots outside the building and window boxes. The house number is 69 Next to the homes is a modern building complex

Text track in the commercial

The actors (A) and (B) have a two-way conversation.

Here is the written script:

Text: EVITE EL EXCESSO (Don't overdrink).

Actor A: *La pase muy bien (I had a good time).*

Actor B: *Si yo tambien. Te llamo si? (Yes, me too. I'll call you. Yes?)*

Actor A: *Si. Bye (Yes. Bye).*

Actor B: *Bye.*

Actor A. *Jose, JOSE! No quieres pasar a tomar algo? (Do you want to come and drink something?)*

Actor B: *Mira necesito ser muy honesto contigo. Tue eres my especial. No quiero quemar etapas en este historia. La sociedad…te pide que seas macho. Pero yo no. Yo quiero ser un caballero. Yo no busco una mujer para pasarme una noche.*

Table 13.14 Code: gestures

Gesture descriptive	Definition	Example
Seduction	Use of body language	(A) Stands in front of door, looking timid (A) Looks into both of B's eyes (B) Kisses A on cheek and turns to leave (A) Rests hand gently on shoulder of B during gentle kiss (A) Brings left knee in towards right knee, which drops the left side of the body and puts it into a submissive pose (B) Steps up to eyelevel of female and hugs A (A) Grabs rose and holds it towards chest
Manipulation	Using body and face to convince	(A) Looks directly at B innocently (B) Stands below A and looks up at her (B) Drops tone of voice and swallows (B) Looks away sheepishly (B) Smiles devilishly as he leaves A. She is unaware of this smile (B) Runs away after A has gone into house
Challenge	Using body and face to demonstrate challenge	(A) Moves left knee towards right knee. Left side of body drops to one side (B) Rolls eyes upward and clenches teeth when A asks if he wants to come into house

Yo busco una mujer para despertarme con ella todas las mananas. Puede parecerte esto muy tonto, muy anticuado lo se.

(Look, I need to be honest with you. You are very special. I don't want to burn moments in this story. Society asks that I be macho. But me, no. I want to be a gentleman. I don't look for a woman to spend just one night with me. I look for a woman to wake up to morning. This may sound very stupid and old fashioned, I know.)

Actor A: *No, no.*

Actor B: *Tu que dices? estamos juntos en ese barco de paciencia y de respeto? (What do you say? Are we together on this boat called patience and respect?)*

Actor A. *Si.*

Actor B: *Okay.*

Text: *Ya comienza el partido. Camerun vs. Mexico. Tecate logo. #somoscaballeros*

(The game has begun. Everything for soccer. We are Tecate. #WeAreGentlemen)

ANNCR: *Somos caballeros. Caballerros de futbol.*

(We are gentlemen. Gentlemen of football)

Music: Piano (Table 13.15).

Table 13.15 Code: text tracks

Text descriptive	Definition	Example
Reverse roles	Female takes role of male, and male takes role of female	(A) "Do you want to come and drink something?" (B) "Society asks that I be macho. But me, no. I want to be a gentleman." (B) "Don't look for a woman to spend just one night with me. I look for a woman to wake up to every morning."
Not my fault	Uses excuses	"Society asks that I be macho. But me, no. I want to be a gentleman." "This may sound very stupid and old fashioned, I know."
Music	Piano	Romantic piano music

The male B switches roles with the female A. He is saying what the woman would like to hear, with the objective of manipulating the female's opinion. For example, he blames the "macho culture that he is part of". It is a demonstration of how "gentlemen" manipulate females so that they can do what they want. The beer Tecate supports this way of thinking by convincing consumers to be gentleman and yet get what they want with women. Tecate promotes the concept that soccer is more important than women.

Conclusion:

Decoding the signs and cultural codes

In 1972, Barthes wrote about the concept of messages and how consumers read connotative meanings as denotative. This concept applied to the Tecate narrative as seen in the television commercial produces the following connotative meaning: the beer named Tecate is actually a gentleman in the eyes of women. What this signifies is that a metal can, filled with light brown liquid, can actually turn any man into a middle class, light-skinned gentleman that is attractive to white females. This meaning is socially and culturally specific and not a natural association.

In *The Quest for Semiotic Meaning* (2007), Danesi defines a narrative as a "[…] story that is put together to portray reality in a specific way. It is a representation of human events as they are perceived to be related to the passage of time […] It is often difficult […] to determine the boundary line between narrative fact and fiction" (2007: 88). The narrative of the Tecate commercial is constructed as a romantic fantasy with the male trying to escape the 'fantasy time' to be in the 'present time.' The communication presents a story of the manipulation of the female stereotype by gentlemen, all supported by a Mexican beer. The light-skinned actors and expensive location sends out the message that to be a "gentleman" you need to live in a white man's world. This ad creates an attractive message to men of all skin types because it presents a sport that is played by men of all races, and males would rather watch soccer together than be at home with the opposite sex. It is male bonding, sponsored by Tecate.

13.4 Conclusion

My thesis question asks, *What are the visual representations of the power relationships in Mexico's political economy as reflected through the appropriation of advertising?* To answer this question, I investigated the history and political economy of Mexico from before the Europeans arrived to the present day.

Aztecs already operated within a classist structure that the Spanish built upon during colonization. The Spanish created a 'castas' system to categorise the peoples of New Spain into socioeconomic and racial categorizations, and during the 18th century, these were reflected in images called 'Pinturas de Castas' that placed the light-skinned Spanish at the top of the socioeconomic and racial pyramid.

Capitalism gained control of Mexico's political economy through what Morton describes as a 'passive revolution'; through cultural hegemony, the capitalist class established institutions that had a cross-class or corporatist character, such as state-sponsored unions and peasant organizations that were based on a clear class hierarchy under the control of the mostly light-skinned ruling class.

The histories of three current ruling families of European descent:—Servitje, Bailleres and Garza—were examined to explore how the continued cultural hegemony is maintained through their powerful connections, and how myth creation, by way of corporate advertising that contain signs and codes, influences Mexico's social and cultural values.

The branded communications broadcast by these families was decoded through performing a semiotic analysis of television advertising created by their companies Bimbo, Palacio de Hierro and FEMSA. The commercials reinforce the myths of the light-skinned ruling class by persuading the Mexican public to buy branded commodities. As academic Jonathan Bignell wrote in the text book *Media Semiotics*, "[…] to possess the product is to 'buy into' the myth, and to possess some of its social value for ourselves[…]" (2002: 31–78).

The semiotic analysis concluded that all three commercials presented light-skinned actors in middle to upper class levels, which is contrary to the actual Mexican demographic breakdown of 83% darker-skinned population called "Mestizo" (Table 13.1) (CIA) (Fitzpatrick). Through their control of capitalism and communication, the ruling class minority have made their lifestyle and skin colour aspirational for the darker-skinned, lower-class majority. The bourgeoisie class is white, confident and wealthy, and the branded messaging presents viewers with the myth of needing protection from the lower classes; the consumers are offered branded commodities to separate themselves.

I conclude that the cultural hegemony cultivated by the ruling families is unconsciously maintained through messages that portray values that indirectly reinforce their class power. The social and cultural values created over 500 years ago are now so ingrained in Mexican society that they are not questioned, and are part of the construction of modern Mexico. The power representations of Mexico's political economy that are presented through advertising is the cultivated myth of the light-skinned ruling class.

13.5 Visual Project: An Exhibition of Messaging

In the visual component of this research, I want consumers to understand how products get linked to myths, and how advertising normalizes some myths which might just be untrue, and as Bignell states, "Advertising has been critiqued as one of the social institutions which perform this function of naturalizing dominant ideologies in our culture" (Bignell 2002: 31–78). In order to highlight the differences between the ruling classes and the rest of the population, I asked creative workers in the lower classes to design their own interpretation of messages for the brands Barcel, Palacio de Hierro and Tecate. I approached two artists: Don Nico and Juan Carlos from Rotulos Interlonas, who both create communications as a practice, and their advertising messages are directed towards Mexico's urban poor. They were asked to create the advertisements using the same brief and copy.

(A) El Palacio de Hierro.
Soy Totalmente Palacio.
(B) Tecate.
#somoscaballeros
(C) Barcel.
Monster Runners.

I supplied both artisans with black and white logos of each of the brands, a product shot and the text I wanted to appear. I asked that they produce an ad in colour for each brand, and the way they decided to express the advertising message was up to them.

The professional name in Spanish for their practice is 'Rotulo,' and Don Nico is a specialist in painting advertising on walls and banners using acrylic paints. Rotulos Interlonas uses digital technology to create and print their advertisements onto 'lonas' made of plastic. They charged $450 pesos for each advertisement pesos by Don Nico and $180 pesos by Rotulos Interlonas (Figs. 13.24, 13.25, 13.26, 13.27, 13.28, 13.29, 13.30 and 13.31).

Once finished, the 'rotulos' were exhibited on the streets of Mexico City from August 26 to August 30, 2015 through an art practice called 'intervention.' They were placed at or close to stores where consumers would be able to purchase or experience the brands advertised, and the placement was recorded using in situ photography (Figs. 13.32, 13.33, 13.34, 13.35, 13.36 and 13.37).

Fig. 13.24 "The brief." Artisan Don Nico and author Carl W. Jones. Palo Solo, Huixquilucan, Mexico

Fig. 13.25 "Rotulos" are printed at Rotulos Interlonas using digital technology. Palo Solo (village) Jesus del Monte, Huixquilucan, Mexico

Fig. 13.26 "Palacio" Digitally printed on plastic "rotulos". Designed and printed by Juan Carlos (right)

Fig. 13.27 "Monster Runners" Digitally printed on plastic "rotulos". Designed and printed by Juan Carlos (right)

Fig. 13.28 "Tecate" Digitally printed on plastic "rotulos". Designed and printed by Juan Carlos (right)

Fig. 13.29 "Palacio rotulo" Hand painted on material. Designed and painted by Don Nico

Fig. 13.30 "Monster Runners rotulo" Hand painted on material. Designed and painted by Don Nico (right)

Fig. 13.31 "Tecate rotulo" Hand painted on material. Designed and painted by Don Nico

Fig. 13.32 "El Country changarro" with rotulo Monster Runners. Naucalpan, Distrito Federal

Fig. 13.33 "Seven Eleven store" in Colonia Condesa corner of Michoacan and Amsterdam, Distrito Federal. Monster Runners plastic rotulo

Fig. 13.34 "Bistrot" in Colonia Condesa corner of Sonora and Amsterdam, Distrito Federal. "Tecate rotulo"

Fig. 13.35 "Bistrot" in Colonia Condesa corner of Sonora and Amsterdam, Distrito Federal. "Tecate acrylico."

Fig. 13.36 "Palacio de Hierro" store in Colonia Roma corner of Salamanca and Colima, Distrito Federal. "Totalmente plastico" rotulo

Fig. 13.37 "Palacio de Hierro" store in Colonia Roma corner of Salamanca and Colima, Distrito Federal. "Totalmente acrylico" rotulo

The interventions performed in this visual project compared the branded messaging funded by the ruling class to the branded messaging created by the lower social economic class. This demonstrated how the absence of casting and expensive narrative techniques seemed to create a situation where the rotulos depended more on design, and for the most part did not reflect the racism and classism of the more mainstream advertising. What is not included in the images created by the two 'rotulistas' is important, as they solved the communication problem by using visual design to construct the message. This leads to questions such as: Does access to technology aid in the construction of racist and classist communication? Can the lower social economic class be able to compete successfully against the ruling class through advertising messaging? Can a new Mexican myth be created without expensive narrative techniques in advertising communications?

References

Adorno, T., and M. Horkheimer. 2002. *Dialectic of enlightenment*, 94–136. California: Stanford University Press.

Barthes, R. 1972. *Mythologies*. New York: Hill and Wang.

Barthes, R. 1974. *S/Z*. New York: Hill and Wang.

BBC. 2010. http://www.bbc.com/mundo/economia/2010/01/100111_1700_femsa_cerveza_med.shtml. Accessed 8 July 2017.

BBC. 2015. http://news.bbc.co.uk/2/hi/business/8451617.stm. Accessed 8 July 2017.

Bernays, Edward L. 1927. *Propaganda/Edward Bernays; with an introduction by Mark Crispin Miller*. Brooklyn, N.Y.: Ig Publishing, c2005.

Bignell, J. 2002. *Media semiotics: An introduction*, 31–78. Manchester: Manchester University Press. Retrieved from http://www.books.google.com/books?id=MGom6ENJRLkI. Accessed 8 July 2015.

Carrillo Trueba, Cesar. 2009. *El racism en Mexico*. Tercer Milenio. Mexico D.F.: CONCACULT.

Census. https://www12.statcan.gc.ca/census-recensement/2006/ref/rp-guides/ethnic-ethnique-eng.cfm. Accessed 9 July 2017.

Chandler, Daniel. 2001. Semiotic for beginners. http://visual-memory.co.uk/daniel/Documents/S4B/. Accessed July 24 2017.

Choro. 2014. TV. http://www.latinspots.com/site/sp/pieza/detalle/18440/comercial/Choro. Accessed 8 Aug 2017.

CIA. 2015. https://www.cia.gov/library/publications/the-world-factbook/geos/mx.html. Accessed 8 Aug 2017.

CIMA. 2013. http://www.cimacnotiias.com.mx/node/62973. Accessed 8 Aug 2017.

CNN. http://www.cnnexpansion.com/negocios/2014/11/17/los-8-ganadores-de-los-antipremios-a-los-anuncios-misoginos. Accessed 8 Aug 2017.

CONAPRED. 2010. http://www.conapred.org.mx/userfiles/files/Enadis-2010-RG-Accss-002.pdf, 40–45. Accessed 8 July 2017.

Danesi, Marcel. 2002. *Understanding media semiotics*. London: Arnold.

Danesi, Marcel. 2006. *Brands*. New York: Taylor & Francis.

Danesi, Marcel. 2007. *The quest for meaning. A guide to semiotic theory and practice*. Toronto: University of Toronto Press.

Danesi, M. 2013. *Encyclopedia of media and communication*. Toronto: University of Toronto Press.

Femsa.com. 2015. http://www.femsa.com/es/content/¿cómo-participa-femsa-en-el-mercado-de-cerveza. Accessed 10 Aug 2017.

Fitzpatrick, T.B. 1975. Soleil et peau [Sun and skin]. *Journal de Médecine Esthétique* 2: 33–34 (in French).

Forbes. (FEMSA). http://www.forbes.com/profile/eva-gonda-de-rivera/. Accessed 8 Aug 2015.

Forbes. 2012. http://www.forbes.com/profile/daniel-servitje-montull/. Accessed 8 Aug 2015.

Forbes. 2015. http://www.forbes.com/profile/alberto-bailleres-gonzalez/. Accessed 8 Aug 2015.

Foucault, Michel. 1975. *Discipline and punishment*. London: Penguin books.

Fuchs, Christian. 2005. *Dallas Smythe today—The audience commodity, the digital labour debate, marxist political economy and critical theory. Prolegomena to a digital labour theory of value*. TripleC. ISSN 1726-67OX.

Gramsci, Antonio. 1971. *Selections from the Prison notebooks; The study of philosophy*, ed. Quintin Hoare and Geoffrey Nowell Smith. New York: International Publishers.

Gramsci, Antonio. 1992. Buttigieg. In *Prison notebooks*, ed. A. Joseph, 233–238. New York City: Columbia University Press.

Harrison, Calire. 2003. Visual social semiotics: Understanding how still images make meaning. *Technical Communication* 50(1): 42–60.

Hernández-Zárate, S.I., A. Medina-Bojórquez, A.L. López-Tello Santillán, and D. Alcalá-Pérez. 2012. Epidemiología del cáncer de piel en pacientes de la Clínica de Dermato-oncología del Centro Dermatológico Dr. Ladislao de la Pascua. Estudio retrospectivo de los últimos ocho años. *Dermatol Rev Mex* 56(1): 30–37.

Horn, Robert. E. 1999. Information design: Emergence of a new profession. In *Information design*, ed. R. Jacobson, 15–33. Cambridge, MA: The MIT Press.

Hussain, Nazia. 2010. Not just a pretty face: The cultural contexts of beauty in Asia. WPP Atticus Awards, Highly Commended, 19. Warc.com. Accessed 20 Aug 2015.

INAH. 2009. *Museo Nacional de la Historia*. Mexico, D.F.: Chapultepec. https://www.youtube.com/watch?v=xms7bYIHxFk. Accessed 8 Aug 2017.

INEGI. http://www.inegi.org.mx. Accessed 8 Aug 2017.

INEGI. 2009. AMAI. http://www.inegi.org.mx/rne/docs/Pdfs/Mesa4/20/HeribertoLopez.pdf. Accessed 8 Aug 2015.

Jewitt, Carey, and Rumiko, Oyama. 2001. Visual meaning: A social semiotic approach. In *Handbook of visual analysis*, 134–156. ed. T. Can Leeuwen and C. Jewitt. London, UK: Sage Publications.

Jones, Carl. 2006. *El show de Guada y lupe*. Colonia Roma, Mexico City: Galeria Films. https://vimeo.com/119641814. Accessed 8 Aug 2017.

León-Portilla, M. 1992. *The broken spears: The Aztec accounts of the conquest of Mexico*, 3–11. Boston: Beacon Press.

Lindlof, Thomas R. 2011. *Qualitative communication research methods*. Thousand Oaks: Sage Publications.

Lizcano Fernández, Francisco. 2005. *Composición Étnica de las Tres Áreas Culturales del Continente Americano al Comienzo del Siglo XXI*. Convergencia. Revista de Ciencias Sociales, num. Mayo-Agosto, Academic investigation (in Spanish), 196; 185–232. Mexico City: University of the State of Mexico.

Marx, Karl, and Friedrich, Engels. 2006. The ruling class and the ruling ideas. In *Media and cultural studies: keyworks*, 9, ed. Meenakshi Gigi Durham and Douglas M. Kellner. Malden: Blackwell Publishing.

Miller, Jacques-Alain. 1977. Suture (elements of the logic of the signifier). *Screen* 18(4) [The original French publication of this article was in 1966.].

MIRA. 2015. *Newspaper Mexico City*. Published Monday 3 Aug 2015. Edition 23.

Morton, Adam David. 2011. *Revolution and state in modern Mexico: The political economy of uneven development*. Lanham, MD: Rowman & Littlefield.

Mulvey, Laura. 1975. Visual pleasure and narrative cinema, 6–18. Originally Published Screen 16.3 Autumn 1975. https://wiki.brown.edu/confluence/display/MarkTribe/Visual+Pleasure+and+Narrative+Cinema.

Palacio. 2015. www.elpalaciodehierro.com Accessed 21 July 2017.

Patterson, Jorge Zepeda. 2007. *Los amos de Mexico.* Mexico D.F.: Editorial Planeta Mexicana.

Sachdeva, S. 2009. Fitzpatrick skin typing: Applications in dermatology. *Indian Journal of Dermatology, Venereology, and Leprology* 75: 93–96. http://www.bioline.org.br.ezproxy.lib.ryerson.ca/pdf?dv09029. Accessed 8 Aug 2015.

Shaheed, Farida. 2014. United Nations Special Rapporteur in the field of cultural rights, 900. http://daccess-dds-ny.un.org/doc/UNDOC/GEN/N14/499/88/PDF/N1449988.pdf?OpenElement, http://ap.ohchr.org/documents/dpage_e.aspx?si=A/69/286. Accessed July 21 2017.

Smythe, D.W. 1977. Communications: Blindspot of western Marxism. *Canadian Journal of Political and Society Theory* 1 (3): 1–28.

Sturken, M., and L. Cartwright. 2009. *Practices of looking: An introduction to visual culture.* New York: Oxford University Press.

Tecate. 2014. http://www.adlatina.com/publicidad/preestreno-de-nómades-para-tecate. Accessed 8 July 2017.

Tecate. http://www.reporteindigo.com/reporte/mexico/no-es-facil-ser-tecate. Accessed 8 July 2017.

Tecate. http://www.reasonwhy.es/actualidad/anunciantes/la-cerveza-tecate-vuelve-al-ataque-con-publicidad-polemica-para-eeuu_2013–04. Accessed 8 July 2015.

Teran/TBWA. (2015) http://latinfashionews.com/2015/02/andresv/. Accessed July 24 2017.

Townsend, Camilla. 2003. Burying the white gods: New perspectives on the conquest of Mexico. *The American Historical Review* 108(3): 659–687.

Triumph of the Will. 1935. https://www.youtube.com/watch?v=GHs2coAzLJ8. Accessed 25 July 2017.

Walker-Smith. 2006. *CBS News Report.* http://www.cbsnews.com/news/cutting-through-advertising-clutter/. Accessed 25 July 2017.

WARC.com. 2008. *The role of helpers in advertising: Bridging the way from semiotics to storytelling.* ed. Joseph Sassoon. Istanbul: ESOMAR. Qualitative Research.

Velencoso, Andres. 2015. http://www.a6cinema.com/en/andres-velencoso-en. Accessed 8 July 2017.

Author Index

A. Olteanu et al. (eds.), *Meanings & Co.*, Numanities - Arts and Humanities in Progress 6, https://doi.org/10.1007/978-3-319-91986-7

Subject Index

A. Olteanu et al. (eds.), *Meanings & Co.*, Numanities - Arts and Humanities in Progress 6, https://doi.org/10.1007/978-3-319-91986-7

Printed by Printforce, the Netherlands